STARDUST TO PLANETS

STARDUST
TO
PLANETS

——

A Geological Tour of the Solar System

Harry Y. McSween, Jr.

ST. MARTIN'S PRESS

New York

Design by Susan Hood

Library of Congress Cataloging-in-Publication Data

McSween, Harry Y.
    Stardust to planets : a geological tour of the solar system / Harry Y. McSween, Jr.
        p.   cm.
    ISBN 0-312-09394-2
    1. Astrogeology—Miscellanea—Popular works. 2. Solar system—Miscellanea—Popular works.   I. Title.
QB455.M35   1993
559.9'2—dc20   93-19056
                                                        CIP

First Edition: August 1993

10   9   8   7   6   5   4   3   2   1

Especially for Lindsay

Contents

Contents

Acknowledgments

The contributions of many generous and talented people have helped to make this book a reality:

To my friends Pat Conroy and Mark Littmann for unflagging encouragement,

To my agent Julian Bach and editors David Sobel and Eric Wybenga for giving a guy a chance,

To my University of Tennessee colleagues Bill Dunne, Ralph Harvey, Ted Labotka, and Rick Williams for red pencil marks and computer wizardry,

To Ed Anders (University of Chicago); Stan Awramik (University of California, Santa Barbara); Tammy Dickinson (NASA Headquarters); Ewe Fink (University of Arizona); Jim Gooding and Mike Zolensky (NASA Johnson Space Center); Mary Ann Hager (Lunar and Planetary Institute); Kathy Hoyt, Baerbel Lucchitta, and Don Wilhelms (U.S. Geological Survey, Astrogeology Branch); Don Lowe (Stanford University); Glenn MacPherson (Smithsonian Museum of Natural History); Ted Maxwell and Jim Zimbelman (Smithsonian Air and Space Museum); Urey van der Woude and Carolyn Young (Jet Propulsion Laboratory); and Brigitte Zanda (Museum National d'Histoire Naturelle) for helping me find stuff,

And to my fellow scientists in planetary geosciences for sharing the thrill of their discoveries.

Thank you.

STARDUST TO PLANETS

The Yellow Pages

―――

Organizing the Solar System

We organize things to get results. Football coaches organize their athletes by physical attributes, assigning the largest to the line and the fleetest to the backfield. Everyone may want to handle the ball, but sorting players by size and skill makes for a winning team. Art exhibits are usually organized around some common theme, such as artist, period, or style. As a result of grouping art in this way, we can discern and enjoy variations on the theme, such as differences in painters' use of color or the evolution of a particular sculptor's style. The Yellow Pages of a telephone directory organizes advertised businesses according to function, so that we can easily find mortgages, morticians, or motels and select from among them the one that meets our needs.

Scientists, too, organize things, as a preliminary step toward understanding them. Butterflies or minerals can be readily classified if you know the rules, but the organization of some aspects of nature is not so straightforward. Classifying the objects that comprise our solar system can be a real challenge.

AN ASTRONOMICAL VIEW OF THE SOLAR SYSTEM

Medieval mathematicians and astronomers deserve much of the credit for the conventional organization of solar system objects. Nicolas Copernicus, in a book published in 1543, the year of his death, proposed that the Sun, rather than the Earth, lies at the center of the solar system. This unorthodox idea was highly controversial, and a Lutheran minister who supervised the book's postmortem publication added an unsigned preface implying that this heretical view was more a convenient calculation scheme than a description of reality. Galileo Galilei was convinced though, and half a century later he expanded the Copernican hypothesis by observing that Venus and Mercury, like the Earth, revolved around the Sun. His telescope also revealed the existence of four moons revolving around Jupiter, thereby demonstrating that centers of motion can in turn be in motion themselves. It was left to Johannes Kepler in 1609 to describe the true nature of the orbits of these bodies. The new order embodied in Kepler's laws of motion finally provided a correct astronomical view of the solar system.

Based on the discoveries of Copernicus, Galileo, and Kepler, astronomers generally organize solar system objects according to their orbital characteristics. We can envision this kind of classification as something like the headings in a Yellow Pages listing: one star, around which revolve nine planets, about some of which revolve a retinue of satellites. To complete the listing, we must include other kinds of bodies that revolve around the Sun but are too small to be classified as planets: asteroids, mostly occupying the otherwise empty space between the orbits of Mars and Jupiter, and comets, lying beyond the outermost planet but occasionally entering the planetary realm. In other words, Ma Bell's organizational listing of the bodies in the solar system would be hierarchiacal and based on their Keplerian orbits. What's wrong with this orderly view of the cosmic neighborhood?

The short answer to this question is: nothing. However, the astronomical view is not the only way to categorize these objects, and fundamental insights often arise from organizing things in different and sometimes unconventional ways. The Yellow Pages recognizes this fact through cross-listings that create several different avenues of approach for finding an advertised business. The purpose of this book is to explore some new and different ways to categorize objects in the solar system.

A GEOLOGIC VIEW OF THE SOLAR SYSTEM

The prevailing doctrine of Copernicus' and Galileo's time held that the Earth stood still. The work of these pioneering astronomers effectively placed the Earth and other bodies in motion, but did nothing to make them interesting objects in their own right. Even today, many people are accustomed to thinking of these objects as immutable—unchanged since the beginning of time. This perception stems in part from the brevity of our human observations, but it is influenced also perhaps by the rather murky views of these faraway bodies that we get from our vantage point on the Earth. But a new truth is emerging, wrought by stunning images gathered by electronic eyes and ears installed on spacecraft and by Earth-based studies of extraterrestrial materials. This revelation is that the Sun, planets, satellites, asteroids, and comets have evolved over time, in some ways ponderously slowly and in others surprisingly rapidly. The solid bodies have been progressively compacted and baked from within, pummeled and sunburned from without; even the stellar centerpiece of our solar system has altered its composition and luminosity. In effect, planets, satellites, asteroids, and comets have been transformed from static points of light into complex worlds shaped by dynamic processes. Such processes can best be described as "geologic," because they are the same as or similar to those processes that have shaped the Earth. As you will see, a geologic perspective has even contributed to our understanding of the one truly astronomical object left in close proximity, the Sun.

Let's return to the idea that categorizing objects from different perspectives can alter our view of the solar system. The astronomical view implies that planets, by virtue of revolving around the Sun, are somehow more important than satellites, which twist less conspicuously around the planets. Let us, for the moment, dispense with pigeonholing solar system objects according to their orbits, and consider a few other characteristics that might serve to categorize them.

We might decide that the *size* of a solar system object is an indication of its relative importance. The differences in planetary sizes are startling. Using this scheme the inner, so-called terrestrial planets— Mercury, Venus, Earth, and Mars—shrink to insignificance beside the outer, or "giant" planets—Jupiter, Uranus, and Neptune. If we include satellites in our inventory, we find another surprise. Ranking objects using some measure of size, say object diameter, forces a juggling of the cosmic pecking order. For example, Jupiter's moon Ganymede and Saturn's moon Titan are larger than either of the planets Mercury or Pluto. Should we arbitrarily define important objects as those with diameters greater than a thousand kilometers, some twenty-five bodies would rise to prominence in this new order, rather different from the nine planets we normally consider important.

Alternatively, we might decide to categorize bodies in the solar system according to their *compositions*. This scheme would divide the solar system basically into two parts. The terrestrial planet region contains worlds comprised mostly of silicate rock with central cores of metal. Conspicuously absent in this region are significant quantities of gaseous elements such as hydrogen and helium that make up the great bulk of the Sun, even though the majority of these bodies have tenuous gaseous envelopes and one even has oceans of liquid water. In contrast, the outer reaches of the solar system are home to planets made mostly of liquids with rocky interiors, sloshing colossi cored by hidden Earths. The liquids were formed by compressing hydrogen and helium gas and ices of water, methane, and ammonia, all simple compounds of the most abundant elements in the Sun. Similar ices still exist in the satellites that surround these behemoth planets. The giant planets thus have bulk chemical compositions much more like the Sun than do the terrestrial planets.

The asteroid belt, marking the boundary between these two compositional terrains, contains both kinds of matter. Asteroids nearer to the Sun have rocky and metallic compositions, whereas those farther out contain both ice and rock. Like the snow line on a mountaintop, this break must correspond to some critical temperature below which vapor could condense as ice in the early solar system.

The solid matter of the solar system originally formed out of a cocoon of hot gas and dust surrounding the infant Sun. This so-called solar nebula was obviously cooler farther away from the fledgling star.

The compositions of planets and asteroids, both of which formed as this solid matter clumped together, are fossil remnants of this primordial temperature variation. The two-tiered distribution of planet compositions that we now see becomes garbled, however, if we consider the satellites gathered about the giant planets. The compositions of these clusters of moons tend to become more icy and less rocky with increasing distance from their host planets, as if they were miniature solar systems. The growing planets somehow must have modified the temperature of the nebula in their immediate neighborhoods.

Another, perhaps more geologic, way of categorizing solar system bodies is by their *densities*. This parameter incorporates information on both of the ways of categorizing planets we have just considered: Density is defined as the ratio of mass (a function of an object's composition) to volume (a function of object diameter). The density of matter is not constant within a single planet or satellite, because the various constituents are not distributed uniformly in its interior. Geologic processes tend to partition the elements into either central cores or surrounding concentric shells, each of which has its own characteristic density. For example, iron, a very heavy element that tends to pack tightly into metal structures, is concentrated in cores, so cored planets are more dense in the middle than near the outside. For this reason, we normally speak of an object's *mean density*, an average value for the entire body.

Mass and volume for planets vary in a decidedly nonintuitive manner. It may seem reasonable to think that as planets become more massive, they should grow proportionately in size. In reality, the increase in volume due to adding more material is mostly offset by more efficient packing of atoms in the highly compressed interiors of large planets. For this reason, Jupiter is only slightly larger in diameter than Saturn, although it is three times more massive.

Not surprisingly, when planets are categorized by density, they break into two distinct groups: the terrestrial and the giant planets. The mean densities of the terrestrial planets range from four to five and a half grams per cubic centimeter, somewhat denser than rocks in your yard. These values are consistent with the idea that such bodies are made of mixtures of rock and metal compacted at high pressures. Most of the giant planets have much lower mean densities of between one and two grams per cubic centimeter, values intermediate between liquid water and rock. This density range is appropriate for worlds made mostly of highly compressed liquids with rocky cores. Saturn, with a mean density of less than one gram per cubic centimeter, falls outside this range. Saturn would float in water, if you could find a big enough bathtub. Much of the planet must be gaseous or icy. (Ice is less dense than water, which is why it floats.) The Earth's Moon has a mean density slightly greater than three, similar to but somewhat less than the terrestrial planets. Most satellites of the giant planets have densities appropriate for mixtures of ice and rock, but Jupiter's inner moons Io and Callisto would have to be grouped with the Earth's Moon in a classification based on density.

There are still other ways of characterizing solar system bodies, but indirect measurements or observations are necessary to gather the required information. As an example, let us say that we wish to categorize planets by their *shapes*. To the casual observer all planets are round balls, but in actuality they are not really spherical. Subtle variations in planetary shape are often easier to measure from the planet's gravitational effects on nearby objects than from visual inspection. A perfectly spherical planet will act as if all of its mass is concentrated precisely at its center. In such a case, the orbits of adjacent moons or spacecraft are predicted exactly from Kepler's laws of motion. In a rotating planet, however, centrifugal forces cause the body to flatten, producing bulges at the equator. Other, more delicate distortions can occur as well, all of which when summed together can mold a planet's overall shape into something like a Japanese lantern. A spinning body may be "out of round" by only a percent or so of its average diameter, but that is enough to induce a corresponding distortion in the planet's gravitational field, which in turn is detectable from its effect on the orbits of nearby objects. Because aircraft and missiles are guided by signals from satellites whose orbits are controlled by the shape of the Earth, knowledge of the precise shape of our own planet qualifies as a military secret.

Shape may seem like a peculiar way to categorize planets, but it carries with it new insights. Shape distortions are sensitive functions of the way that materials of different density are distributed inside the planet. By comparing the sizes of the bulges to the rotation rate that induced them, we can better understand compositional and density variations in its interior. For example, we have learned from analyzing its shape that Saturn, surprisingly, has a larger rocky central core than does Jupiter. If we could correct for the different rotation rates of the various solar system bodies, a classification based on shapes would reveal a new hierarchy of objects, reflecting the sizes and compositions of their hidden, dense cores.

Another means of categorizing objects is by taking their *temperatures*. Any body of substantial size generates heat deep in its interior, but we can measure this heat only as it escapes at the surface. The fevers of the terrestrial planets are caused by gradual breakdown of unstable atoms, called radioactive isotopes, into smaller elemental particles. In their transmutation, the radioactive isotopes give off heat as a by-product. This heat gradually makes its way toward the planet's surface, ultimately escaping to space. Heat flowing from the Earth's interior must be measured by lowering a temperature probe into a hole in the ground, because the internal heat is overwhelmed by the warming effects of sunlight once it reaches the surface. You might correctly predict, then, that it would be virtually impossible for spacecraft instruments to detect the heat emanating from planets near the Sun. Nevertheless, we can gain some insights into a planet's radioactive furnace by observing geological manifestations of internal heating, such as active volcanic eruptions and crustal movements. Conversely, dead volcanoes and tectonic inactivity tell us that a planet's internal heat engine has cooled.

The smaller amounts of sunlight reaching the outer planets permit the warmth flowing from their interiors to be measured directly by spacecraft instruments. After correcting such measurements for modest amounts of solar heating, planetary scientists have discovered that Jupiter, Saturn, and Neptune all give off substantially more heat than they receive from the Sun. Why Uranus does not radiate more internally generated heat remains a puzzle. Radioactive decay in the giant planets must produce heat, just as in the terrestrial planets; however, these planets are far too warm for the observed outflows of heat to be explained by this mechanism alone. How then can we explain this observation?

Another way to produce heat is by compression. If you have ever used a bicycle pump to inflate a tire, you may have noticed that the pump barrel got warm as the air inside was compressed. Liquids within the interiors of the giant planets are compressed by the pull of gravity on the overlying matter, producing heat in an analogous manner.

The recognition of planetary compressional heating leads to a fascinating insight into the relationship between large planets and stars. We tend to think of planets and stars as distinct *kinds* of matter, when in actuality they are merely different *amounts* of matter. The early Sun, like the giant planets, was heated by compression, as it grew in size by addition of gas and dust from the solar nebula. But compression of the much larger proportion of matter in the Sun ultimately allowed it to reach the 10 million-degree threshold required to initiate thermonuclear fusion reactions. So, Jupiter and its relatives behave something like stars before their thermonuclear furnaces are ignited. This analogy between giant planets and stars is not really that farfetched. Orbiting other stars, there may well be planets even more massive than Jupiter, superplanets that astronomers call brown dwarfs. Such objects theoretically can have masses up to eighty times that of Jupiter, so they are dwarfs only in comparison to stars. Because compression increases rapidly as mass is added, some of these massive objects actually may be packed into smaller volumes than Jupiter. Such bodies are not massive enough to trigger thermonuclear fusion, but they should give off substantial amounts of heat from compression—hence the reference to brown, or dim starlike objects. Unfortunately, brown dwarfs are probably too dim to be seen directly by telescope, but searches are under way to discover the existence of such objects by their gravitational effects on nearby stars or by their faint infrared emissions of heat.

But let's stick to what is actually in our own solar system. Categorization of solar system bodies by temperature is tantamount to assessing how geologically active they are. The largest terrestrial planets have the hottest interiors, because the heat gradually generated by radioactive decay takes longer to migrate out of a larger body. Earth and possibly Venus are volcanically and tectonically active, whereas the other smaller planets and the Earth's Moon are not. We do not yet understand enough about heat production in the giant planets to make blanket statements. Temperatures within a few of the planet-size moons of the giant planets offer a surprise, but let's save that for a later chapter.

From these examples, it should be obvious that different planets and moons, and even the asteroids and comets that we have not yet considered, have their own individualized geologic histories, which may not relate at all to their astronomical status. In fact, there is mounting evidence that an object's orbital identity need not remain fixed for all time. The orbits of some solar system bodies appear to have changed to the point where they acquired new astronomical status. Phobos and Deimos, the tiny moons of Mars, are good examples. These former asteroids were somehow snagged from orbits around the Sun during close approaches to the planet Mars. Likewise, the planet Pluto may once have been a Neptunian moon, perhaps perturbed into a planetary orbit by some massive collision.

THE ORGANIZATION OF THIS BOOK

Most books about the solar system march outward from the Sun, planet by planet and moon by moon, in lockstep with astronomical ordering by orbital position. At each encounter, the body is described in its entirety, a systematic grand tour extending to the edge of interstellar space. In this book I have adopted a very different tack. My intent is not to provide a comprehensive survey of solar system objects, but to illustrate how a geologic perspective can provide a new understanding of our cosmic neighborhood. I will do this by using a few carefully selected examples. We will jump about almost at random, from the broiling surface of Venus to the icy nucleus of Comet Halley, from tiny motes of stardust in meteorites to ancient river channels on Mars. At each stop, we will focus on some geologic materials, phenomena, or processes that make these objects more than just pinpricks of light in the night sky.

In so doing, we will unfortunately bypass reams of important data on planets, satellites, asteroids, and comets. A common practice in books about the solar system is to tabulate all the physical, compositional, and orbital characteristics of these objects. This practice is, of course, the most efficient (but also the most boring) way to transmit a lot of hard-won information. Rather than compile tables of data, at the end of each chapter I have included an annotated bibliography of especially good readings that will provide additional detailed information for those who are interested.

This book does not even pretend to provide a complete Yellow Pages listing for the solar system. However, as your fingers do the walking through the following sections, you will learn about many exciting discoveries and, I hope, gain fresh perspectives into the origin and geologic evolution of the solar system around us.

Some Suggestions for Further Reading

Beatty, J. K., and Chaikin, A., eds. 1990. *The New Solar System*, 3rd ed. Cambridge, MA: Sky Publishing Corporation-Cambridge University Press. This book contains twenty-three highly informative contributions on solar system science by experts in the field; it is perhaps the best single reference on the solar system available to nonspecialists. Its appendixes include tables of data on planets, satellites, asteroids, and comets as well as a collection of planetary maps.

Gore, R. 1985. "The Planets: Between Fire and Ice." *National Geographic*, vol. 167, no. 1, pp. 4–51. A very readable overview of the planets, gloriously illustrated as you might expect from this magazine.

Hamblin, W. K., and Christiansen, E. H. 1990. *Exploring the Planets*. New York: Macmillan Publishing Company. An excellent introductory textbook containing much up-to-date information on solar system objects.

Moore, P. 1988. *The New Atlas of the Universe*. New York: Arch Cape Press. Contains stunning photographs and illustrations of planets and satellites; the last third of the book focuses on topics outside the solar system.

Morrison, D., and Owen, T. 1988. *The Planetary System*. Reading, MA: Addison-Wesley. This easily understandable text provides a wealth of information on planetary bodies.

Preiss, B., ed. 1985. *The Planets*. New York: Bantam Books. An interesting set of paired science and science fiction articles by noted researchers and writers; some of the science articles are now somewhat dated because of spacecraft encounters with planets since 1985.

Taylor, S. R. 1992. *Solar System Evolution: A New Perspective*. New York: Cambridge University Press. This is an engrossing account of the origin of the planets and the processes that have subsequently modified them.

Watching the Man
in the Moon

A Lunar Stratigraphic Time Scale

One of my favorite sports figures is Yogi Berra, the legendary catcher for the New York Yankees. Yogi has parlayed his wealth of experience as a baseball player into a second career as a coach and manager. In this role he is interviewed by the media, and nowadays he is probably best known for his occasionally confused but always colorful phrasing. One example from among his many memorable quotations is "You can observe a lot just by watching." This statement is perhaps more profound than it may seem on the surface. When used in a context that I am sure Yogi never intended, it describes the empirical method of scientific exploration. As translated, Yogi's axiom means that many scientific discoveries are made simply by watching the natural world, rather than by experimenting with it.

Certainly a lot of rather fundamental contributions to planetary geology have resulted just from these kinds of observations. A good example of the application of the Yogi method is in the construction of a geologic time scale for the Moon. This brambled story unfolded painfully slowly as generations of Earthbound scientists watched the Moon and the Earth around them, building on the observations of their predecessors and gradually learning from their insights.

SOME EARLY LUNAR OBSERVATIONS

This tale began halfway through the seventeenth century in the Baltic seaport of Danzig, now known as Gdansk, Poland. Johann Holwelcke, a rich merchant, city councillor, and part-time astronomer, produced the first map of the Moon that is reasonably accurate by modern standards. In 1647 Holwelcke, with the able assistance of his wife Elizabeth, published his *Selenographia*. As was the custom for scientists of his time, Holwelcke Latinized his name to Johannes Hevelius. He was not by any means the first person to attempt a lunar map, but, unlike his predecessors, Hevelius rather faithfully recorded many of the details we see in modern photographs. He was a skillful observer, and his lunar map stood as the best available for over a century.

Hevelius built an observatory on the roof of his home and equipped it with the best telescope available at that time. Because the two convex lenses in a telescope produce an inverted image, we can infer that Hevelius actually constructed his map upside down. However, he had the good sense to publish his map with north at the top, in conformity with the familiar view of the Moon we see with the unaided eye. The orientation of lunar maps has remained a point of contention throughout the ensuing centuries. Astronomers have generally felt more comfortable with inverted maps, like the images in their telescopes, whereas planetary cartographers and geologists have fought to restore north to its rightful position at the top. If you don't think it matters, try looking at a map of some familiar terrestrial spot upside down.

It is a common misconception that the Moon always presents exactly the same face to the Earth. Although it is true that the Moon's rotation is synchronized with its revolution about the Earth, periodic wobbles, called librations, in the Moon's rotation intermittently reveal other parts to Earthbound watchers. Hevelius was the first to examine carefully these edges, or limbs, which he illustrated on his map.

Hevelius' map nicely illustrates the contrasting light- and dark-colored areas that comprise the lunar surface. These distinctive terrains, which in Hevelius' time were named *terrae* (lands) and *maria* (seas) respectively, are visible with the naked eye and form the basis for the familiar "man in the Moon" (or the less familiar "woman in the Moon" from Hawaiian legend). The same terms are still used

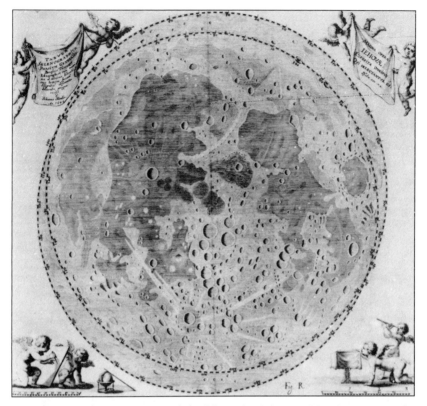

This seventeenth-century map of the Moon, by Johannes Hevelius, accurately portrays a number of features now recognized in modern lunar maps. Hevelius also attempted to map portions of the Moon's peripheral regions that are observable during periodic wobbles (the areas between the inner and outer dashed arcs). A banner carried by cherubs in the upper left corner of the map contains a Latin inscription describing the work and identifying its authorship (although no mention was made of his wife Elizabeth's collaboration), and another in the upper right contains a biblical quotation.

today, although we now recognize that the terrae are not at all like continents on the Earth and that the maria are dry as a bone. The *Selenographia* also shows the locations and relative sizes of several hundred circular craters, as well as systems of bright rays emanating from Tycho, the large crater in the lower hemisphere, and Copernicus, a large crater situated on the maria just above the equator. Hevelius actually named all these features after what he thought were terrestrial counterparts; for example, Copernicus crater was called Etna, after the Italian volcano. A few years after Hevelius published his map, the convention of naming lunar features after scientists who had studied the Moon came into general use and his nomenclature was abandoned. Hevelius correctly illustrated the boundaries of the dark maria and recognized subtle variations in brightness within the maria, although he did not record them very accurately.

STRATIGRAPHIC PRINCIPLES

While Hevelius' map illustrates many geologic features that could have been used to construct a lunar time scale, such an exercise was well beyond his ability. It is probably safe to say that the possibility never even occurred to him. This is not meant as disrespect for Hevelius—the critical observations that were necessary for the construction of a planetary time scale would not be made for another several decades. In fact, the one individual who might have helped Hevelius at this juncture was only nine years old and living in Denmark at the time the map was published. This lad, Nils Steensen, was probably more focused on biological pursuits; in time, he grew up to become the court physician to the Grand Duke of Tuscany in Italy. But despite Steenson's obvious success in the medical profession, his heart just wasn't in it. Instead, he preferred to putter around in the solitude of the northern Italian countryside, watching rocks rather than tending to cuts and bruises. His patron appears to have been remarkably understanding, or at least healthy, freeing Steenson to spend ample time away from the court making observations of layered geologic units, or "strata." At some point Steenson Latinized his name to Nicolaus Steno, a name that is now universally recognized by geologists.

Steno is credited with the formulation of many of the basic principles by which the geologic record is interpreted. Two of these are applicable to understanding lunar geology.[1] Steno's first contribution, the principle of *superposition*, can be understood from the observation

[1] *Geology* literally means "study of the Earth." *Lunar geology* is thus somewhat of a misnomer, but it is probably the best we can do. The alternative—*selenology*—is cumbersome, and the use of latinized "-ology" terms for each planet or satellite, as in *Venusiology*, would reduce the study of solar system objects to an obstacle course in names. The term astrogeology is sometimes applied to the study of extraterrestrial bodies.

that a stack of pancakes must be built from the bottom up, rather than from the top down. Applying this idea to stratigraphy, the principle means that in any sequence of undisturbed strata, the oldest unit lies at the bottom and successively higher layers are progressively younger. This may not appear to be something worth writing home about, as it is intuitively obvious. In fact, this principle probably was already understood by a number of Steno's fellow naturalists, but he was the first to explain the concept formally. In fairness I should also note that applying this principle to deformed strata can pose somewhat of a challenge to the geologist.

The second of Steno's utilitarian concepts is equally straightforward. The principle of *original lateral continuity* tells us something about how horizontal strata end. A spoonful of pancake batter dropped into a frying pan will spread out, either thinning at its edges or abutting against the edge of the griddle or another pancake. Likewise, a layer of geologic strata originally must extend horizontally in all directions until it gradually thins to a feather edge, ends abruptly against some barrier to deposition, or grades laterally into some other kind of material. This principle provides quite a few possibilities, and Steno seems to be covering all bases here. The point is, though, that an original stratigraphic layer is not laid down as individual segments isolated from each other; rather, it forms as a continuous sheet. This is the principle that leads the geologist to examine opposite walls of a canyon to match up particular stratigraphic units as parts of the same original layer that have been dissected over time by river erosion. Using these ideas, it is possible to understand the order in which strata were deposited or, in other words, to begin to erect a stratigraphic time scale.

Steno's stratigraphic principles were published in 1669. Although Hevelius' lunar map was published decades earlier and may have been available to Steno, his attention seems to have been riveted at his feet rather than toward the heavens. Well over two centuries would pass before anyone noticed that these principles could be used to decipher the geologic history of the Moon. This does not mean, of course, that the geologic community was twiddling its thumbs during this period. Many of the soon-to-be giants of this discipline were busy during the eighteenth and nineteenth centuries, providing a global framework within which Steno's principles could be used to understand the Earth's geologic history. During this interval battles were fought over interpretations of the rates of geologic processes, from which emerged our current appreciation of the immensity of geologic time.

Also added during this period was yet another important concept, known as the principle of *crosscutting relationships*. Returning to our pancake analogy, the idea behind this new principle is that you have to make a pancake before you can slice it into pieces. In other words, any geologic feature that cuts across a rock unit must be younger than the unit that is crosscut. Crosscutting geologic features could be erosional surfaces where material has been stripped away, or faults along which juxtaposed rock units have slipped. Again, this concept is a statement of the obvious, yet its application to real rocks can be challenging.

But let us pretend that Steno or one of his Victorian successors did make the connection between these stratigraphic principles and the features shown on Hevelius' lunar map. What could have been learned about lunar geologic history at that point? Based on the concept of crosscutting relationships, he could have inferred that the visible lunar craters formed later than the layers they pockmark. This could have been said with confidence, even though exactly how the craters formed may have been a mystery. It is not possible to tell from Hevelius' map whether the dark maria regions overlie or underlie the bright, heavily cratered terrain. Steno might have been tempted to assume that the isolated patches of maria were originally one unit. Applying the principle of original lateral continuity would then have led him to the conclusion that the terrae are younger. As we will see later, this particular conclusion would have gotten him into trouble. Finally, he could have discerned clearly that the bright rays radiating from Tycho and Copernicus are superimposed on the surrounding units, both terrae and maria. The principle of superposition then dictates that these bright rays formed late in lunar history. If these rays formed at the same time as the craters from which they emanate, then the craters too must be rather young. Thus we can see that the stratigraphic principles formulated by Steno and later geologists provided important tools that Victorian geologists could have used to begin to reconstruct the Moon's geologic history. Unfortunately, no one tried.

Steno and the other scientists who crafted the principles of stratigraphy practiced their art as a hands-on (or, actually, feet-on) activity.

They examined terrestrial strata at close range, even to the point of using the tiny but distinctive fossils contained in these layers to correlate them from one place to another. Perhaps even more important, they were able to see the strata in three dimensions in some places, such as where they were exposed on cliff faces. The necessity of having three-dimensional information derives directly from the principle of superposition: If you are looking down on strata from above, it may be difficult to determine the vertical positions of different layers, so that the relative age relationships between different strata are ambiguous. Plying the stratigrapher's trade is possible in the two-dimensional world of a map, but it is certainly easier when one is actually on the ground to examine the vertical succession of strata. There is no getting around the fact that stratigraphy is a three-dimensional exercise. The trick in applying this methodology to the Moon or planets is inferring the third dimension from a two-dimensional presentation. What was really needed to tweak someone's interest in lunar stratigraphy was a better map or, more accurately, a better image, one that carried three-dimensional information.

PHOTOGRAPHS AS GEOLOGIC TOOLS

Photographs do a much better job of portraying the vertical dimension than do maps. My own sketches could be considered three-dimensional only by the kindest of art critics, but with the aid of a camera I can easily make an image that shows depth of field. What property is it that allows one to visualize the third dimension in photographs? The answer, of course, is shadows. I learned this first-hand on an Antarctic expedition to search for meteorites in late 1980. Our field party spent several days confined to camp because of a phenomenon known as white-out. This condition occurs whenever thick clouds hug the ice sheet, producing a dazzlingly bright fog. The cloud particles diffuse sunlight, so that the light is not coming from any particular direction. The white surface of the ice sheet is commonly rather rough and undulating. Walking on this surface normally poses no problem because the dips and ridges are made readily visible by shadows. In the white-out condition, though, no shadows are cast because the light is nondirectional. The vertical dimension is missing, and the ice surface looks perfectly flat when viewed from a standing position. For the duration of the white-out, I was stumbling around like a drunk and jarring my teeth with almost every step.

Shadows on telescopic photographs also serve to delineate three-dimensional features. Details of vertical features are seen most readily near the *terminator*. This term has acquired a grisly meaning in some recent movies, but in this context it refers to the line of shadow marking lunar sunrise and sunset. Near the terminator, even the subtlest hills and valleys are strongly shadowed. A series of photographs, taken at different times and showing the terminator marching across the face of the Moon, provides just the kind of data set that could excite an Earthbound lunar stratigrapher.

TOWARD A LUNAR STRATIGRAPHIC TIME SCALE

As is often the case in science, there were a few false starts in the development of lunar stratigraphy. Grove Karl Gilbert, another household name in geology, appears to be the first person to contemplate melding stratigraphy and lunar observations. In the late 1800s Gilbert was secure in his position as chief geologist of the United States Geological Survey, so he could work on a zany project like this. He already had made his mark in terrestrial geology, gaining international renown by studying mountain building, glacier movements, and the origins of the Great Lakes and Great Salt Lake. He also was interested in how craters formed on the Moon. Based on his experience with terrestrial landforms and his interest in lunar geology, Gilbert was the logical person to make this conceptual leap. In a paper published in 1893, he described his view of the ejection of material from a giant impact on the Moon (the scar of which we now recognize as the Imbrium basin). The primary focus of this paper was on proving that craters formed by impacts, and almost offhandedly he suggested that the geologic history of the Moon might be constructed around the strata laid down by the Imbrium impact event. The real breakthrough here is the recognition that impact ejecta, like pancake batter that is splattered out of a mixing bowl onto the kitchen counter, should obey the same stratigraphic laws as terrestrial strata deposited by water or wind. As a result of this landmark paper, Gilbert is commonly remembered as the author of the impact interpretation for lunar craters. Unfortunately, he never followed up on the idea of using crater ejecta as stratigraphic markers, and neither did his contemporaries. The concept was resurrected several times in the first half of the twentieth century, but it never generated any flurry of scientific activity. The Moon was so remote—lunar stratigraphy just seemed too much like science fiction.

All this changed in 1962 when two geologists with the United States Geological Survey laid out in specific detail how geologic units of the Moon could be mapped and a stratigraphic sequence comparable to that on the Earth could be defined. Gilbert had anticipated that this was possible, but he did not prove that it would work. Within the short span of twelve published pages embedded obscurely in the middle of a rather dense symposium volume, Eugene Shoemaker and Robert Hackman convincingly demonstrated that the geologic history of a portion of the Moon was decipherable using stratigraphic principles. This time the scientific world paid attention.

Shoemaker and Hackman's paper is a wonderful contribution and fully deserves the recognition it has received, but it didn't hurt their cause that the United States and the Soviet Union were then engaged in a race for the Moon and glory. The importance of lunar geologic maps, demonstrated as feasible by these geologists, in selecting future landing sites was not lost on mission planners from NASA (the National Aeronautics and Space Administration). What may have been even more important in bringing the Moon closer to home at this time, however, was the arrival of photographs taken from spacecraft. Beginning in the summer of 1964, much of the surface of the Moon, frontside and backside, would be photographed at close range by three impacting spacecraft (Rangers), thirteen soft-landing spacecraft (Surveyors and Lunas), fourteen flyby or orbiting satellites (Zonds and Lunar Orbiters), and eight manned spacecraft (Apollos). Before looking at Shoemaker and Hackman's map, let's see what was so special about these new photographs.

Lunar images taken close-up by orbiting spacecraft provided a wealth of detail that was previously unattainable. Features as small as an automobile could be seen in many photographs, with the exact resolution depending on the distance of the camera above the lunar surface. In addition, pictures of the craggy "backside" of the Moon revealed that maria were mostly frontside phenomena, although the reason for this difference was not obvious. Spacecraft photography normally was designed to make maximum use of low Sun angles. In this way, details of the three-dimensional surface topography were accentuated by shadowing. Most pictures were taken at lunar sunrise, with the spacecraft orbiting from west to east. Therefore, the shadows on properly oriented photographs are cast from east to west.

The spacecraft cameras worked much the same as any other kind of camera, but the film was in strips long enough to accommodate hundreds of pictures. The exposed film was then either returned to Earth for developing, as done by the Soviet Zond and manned Apollo missions, or developed on board by pressing it against a mat soaked with developing and fixing solutions. The Lunar Orbiter, the first orbiting spacecraft with the technology to develop its own pictures in flight, was popularly dubbed the "flying drugstore." The image on film processed in flight had to be sent back to Earth. This was done by a scanner, essentially a small beam of light that moved systematically over the film. A detector located on the opposite side of the film recorded the amount of light that was transmitted through the clear portions or blocked by the darker parts. These brightness variations were then turned into video signals for transmission. The scanner moved ceaselessly back and forth across the film in a series of long traverses, called framelets, and the photographic image was assembled piecemeal as these framelets were encoded and sent to Earth. The boundaries of adjacent framelets did not always match precisely, and discontinuities in orbiter photographs can easily be mistaken for linear geologic features. Because photographs taken during the Apollo missions were hand-carried back to Earth, they were free of framelet distortions. Nowadays more sophisticated imaging systems similar to television cameras or facsimile machines are used in spacecraft.

Lunar photographs are dominated by craters of all kinds—big and small ones, craters with peaks in the middle, prominent bowl-shaped craters and subdued depressions. Craters are such fascinating features that Moon watchers have historically focused on their form and origin almost to the exclusion of other aspects of lunar geology. Understanding how craters form is a prerequisite to developing a lunar stratigraphy. At the time Shoemaker and Hackman's paper was published, argument still raged about whether craters were volcanic depressions or the scars of impacting meteors, despite Gilbert's valiant effort to win converts to the impact hypothesis. The preponderance of scientific opinion, however, was clearly swinging to the impact

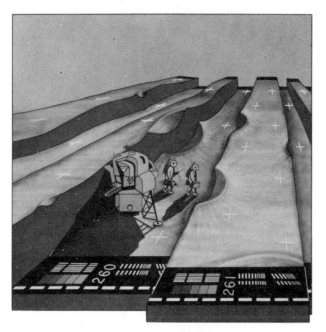

Well . . . . I'll be damned !

Before their missions, the Apollo astronauts got an ample dose of lunar orbiter photographs full of framelet discontinuities. This cartoon suggests the effect that staring too long at such photographs might have. *(J. W. Van Divier, U.S. Geological Survey.)*

argument, a view that Shoemaker and Hackman accepted. It turns out that the impact origin was the correct one, although it may not have mattered a great deal, as either process could produce strata. We know from studies of terrestrial craters that large impacts pulverize vast quantities of the target rock, ejecting this material outward to leave behind an excavated bowl. More important from a stratigraphic viewpoint, though, is what happens to the ejecta. This material forms a mantle about the crater, in effect a stratigraphic layer deposited almost instantaneously.

Shoemaker and Hackman focused their attention on a portion of the Moon near the crater Copernicus. From lunar photographs, they identified distinctive rock units and constructed a map of the Copernicus region. Some of these units were the splattered blankets of material ejected from large craters, and others were lava flows. They then applied the principles of superposition and crosscutting relationships

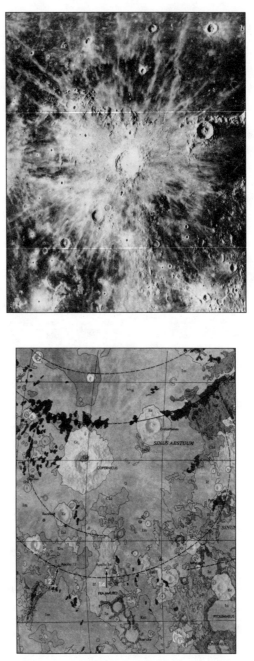

This photograph of the Copernicus region illustrates how the ejecta blanket of this young crater has been superimposed over Imbrium ejecta deposits and the dark mare lavas. Crater Eratosthenes, another important and somewhat older stratigraphic marker, occurs to the northeast. Also shown is a geologic map of this region, modified from the now-classic map published by Shoemaker and Hackmann. *(Mount Wilson Observatory and U.S. Geological Survey.)*

to these strata, carefully documenting which layer of ejecta was super-imposed on which crater and vice versa. After a time they were able to distinguish five units that formed at different times. The upper four deposits—the Imbrian, Procellarian, Eratosthenian, and Coperni-can—were each named for a specific impact or volcanic event. The lowermost and therefore oldest deposit, called Pre-Imbrian, is actually a composite unit consisting of anything deposited before the Imbrium event. Shoemaker and Hackman's paper hinted at the possibility that the Pre-Imbrian unit might be subdivided after it had been studied in more extensive exposures elsewhere on the Moon, and that is just what happened.

The Imbrian deposit, as defined by these geologists, is a widespread blanket of ejecta from an enormous impact that formed the Imbrium basin, the largest crater on the Moon. Most of the dark maria regions were lumped together into Procellarian strata, named for Oceanus Procellarum, the largest of the maria. Shoemaker and Hackman cor-rectly interpreted these as flows and lakes of lava that had pooled in older craters like Imbrium. They divided the craters superimposed on, and thus younger than, the maria into two distinct units. The older Eratosthenian deposits, named for the crater Eratosthenes, consist of craters and ejecta blankets without bright rays. Craters like Coperni-cus and Tycho that do have bright rays were assigned to the Coperni-can unit, and are the youngest observed features on the Moon.

As other geologists attempted to map the rest of the Moon using the same stratigraphic principles, they found it necessary to modify Shoemaker and Hackman's sequence of units. This kind of adjustment is commonplace in terrestrial stratigraphy too. It is not merely tinker-ing with nomenclature; rather, it is often a redefinition of units forced by the accumulation of new information from other areas. In the case of the Moon's stratigraphy, the first change was to downgrade the Procellarian lava unit, lumping it as part of the Imbrian deposits because it formed at nearly the same time. The other major change was a subdivision of the Pre-Imbrian into two sets of deposits—Nectarian and Pre-Nectarian—depending on whether they formed before or during the excavation of the Nectaris basin, a large, subdued crater on the eastern side of the Moon. Nectarian deposits are ejecta from this ancient crater. The Pre-Nectarian unit predated the Nectaris basin and constitutes the oldest terrain on the Moon.

TOWARD AN ABSOLUTE TIME SCALE

The stratigraphy worked out by Shoemaker, Hackman, and other geologists identified a series of important events on the Moon, and their ages *relative to each other* were known. The major challenge remaining was to put real dates on these events. This had to await the return of lunar samples by Apollo astronauts. The ages of lunar rocks, determined from measurements of the decay of radioactive isotopes in them, placed limits on these events. Once each analyzed lunar sample was assigned to a particular stratigraphic unit, the age of that unit was then known. For example, many *Apollo 14* samples are ejecta from the Imbrium basin, so their 3.8 billion-year-old ages are thought to mark the time that the crater formed.

Samples returned to Earth by the Apollo missions were collected from a mere six locations on the frontside of the Moon, most of which were selected to maximize landing safety rather than scientific payoff. A cynic might say that inferring the ages of important geologic events based on lunar rocks from only a handful of sites is like trying to determine the geologic history of the Earth from gravel in a few places such as Kuwait, Ecuador, and Nova Scotia. This is not a fair criticism, though, because the ages determined for lunar rocks have been supplemented with other age data. The ages of unsampled areas of the Moon can be estimated by an ingenious technique called crater counting. The longer a surface has been exposed to bombardment by meteors, the more craters it will have. By counting the number of craters on a particular stratigraphic deposit, it should be possible to estimate its age. Of course, there are a few problems with this idea. The cascading debris from large impacts produces smaller, secondary craters, so the crater counts must be adjusted to remove secondary craters. Also, the rate of crater formation was much greater in the distant past, but this variation can be assessed by counting craters on different lunar strata, the ages of which have been determined previously by isotopic dating of returned rocks. Another potential problem is that very old surfaces may become saturated with craters, so that older craters are obliterated by newer ones. In this situation, the number of craters and thus the apparent age of the unit remain constant. Despite these difficulties and uncertainties, crater counting can provide some information on the ages of lunar geologic units that

This series of paintings illustrates the geologic evolution of the Moon, as inferred from lunar stratigraphy. In the first view, just after the Imbrium impact, the ancient surface is heavily cratered and the Imbrium basin is prominent. The appearance of the Moon changed dramatically with the subsequent eruption of mare lavas, as shown in the second view. Crater Eratosthenes had not yet formed. The present Moon, illustrated in the third view, shows the effects of later impacts. Eratosthenes and similar craters without bright rays formed first, followed later by bright-rayed craters such as Copernicus. *(Paintings by D. Davis, U.S. Geological Survey.)*

have not been sampled directly, and therefore allow those units to be correlated with other stratigraphic units that formed at the same time. This is particularly helpful when mapping areas far removed from important stratigraphic markers such as the Imbrium ejecta blanket.

Isotopic and crater-counting ages have now been combined with lunar stratigraphy to piece together a rather comprehensive picture of lunar geologic history. The earliest rocks, assigned to the Pre-Nectarian deposits, form heavily cratered units that predated the formation of the Nectaris basin. These rocks formed during the waning stages of the assembly of the Moon from smaller bodies about 4.5 billion years ago and the melting of the outer part of the Moon soon after to form the terrae. A major impact produced the Nectaris basin and its ejecta blanket 3.9 billion years ago. A number of other large craters formed at about the same time, so this must represent a particularly violent period in lunar history. The Imbrian event began 100 million years later with a gigantic meteor impact that excavated the Imbrium basin and scattered its ejecta over a vast area of the lunar nearside. Before the Imbrian period was over at about 3.2 billion years ago, most of the lavas that compose the maria would erupt and solidify. Extrusion of the remaining maria lavas appears to have overlapped the later Eratosthenian crater-forming events. The final episode of lunar history is recorded by the Copernican deposits, which include craters with bright rays that formed no more than about 1.1 billion years ago. The present Moon shows the ravages of all these integrated events.

The stratigraphic approach has proved to be just as utilitarian in lunar geologic exploration as it has been in unraveling the geologic evolution of the Earth. Where before the face of the Moon was just a profusion of named craters, maria, and other random features, it now has an orderly geologic history. And we observed all this just by watching. Yogi was right.

Some Suggestions for Further Reading

Carr, M. H., Saunders, R. S., Strom, R. G., and Wilhelms, D. E. 1984. *The Geology of the Terrestrial Planets.* NASA Special Publication 469. Chapter 6 by Wilhelms gives a superb and comprehensive description of lunar geology and stratigraphy, and the colored maps at the end of this section summarize the Moon's geologic evolution especially well.

El-Baz, F. 1980. *Gilbert and the Moon*. U.S. Geological Survey Special Paper 183, pp. 69–80. A historical account of G. K. Gilbert's pioneering work in lunar geology.

Greeley, R. 1987. *Planetary Landscapes*. Winchester, MA: Allen and Unwin. Chapter 4 of this textbook gives a beautifully illustrated description of the general physiography and geologic history of the Moon.

Greeley, R., and Batson, R. M., eds. 1990. *Planetary Mapping*. Cambridge, England: Cambridge University Press. This excellent book provides a wealth of information on how planetary maps are constructed, how various features are named, and how they are used to interpret a planet's geologic history. Chapter 2, by R. M. Batson, E. A. Whitaker, and D. E. Wilhelms, is an especially good summary of the history of planetary cartography.

Mutch, T. A. 1970. *Geology of the Moon—A Stratigraphic View*. Princeton, NJ: Princeton University Press. A pre-Apollo treatment of lunar geology, now somewhat dated, but strongly focused on the stratigraphic view.

Shoemaker, E. M., and Hackman, R. J. 1962. "Stratigraphic Basis for a Lunar Time Scale," in Z. Kopal and Z. K. Mikhailov, eds., *The Moon*. London: Academic Press, pp. 289–300. The short but classic paper that demonstrated the fundamentals of lunar stratigraphic mapping.

Wilhelms, D. E. 1987. *The Geologic History of the Moon*. U.S. Geological Survey Professional Paper 1348. An exhaustive treatment of lunar geology, including lucid descriptions of current lunar stratigraphy.

Could You Eat a Comet?

A Close Encounter with the
Nucleus of Comet Halley

Growing up in South Carolina, I hardly ever saw snow. Perhaps once every few years several inches might fall, only to disappear within hours. As you might expect, the term snowball does not conger up childhood memories of heaving packed white projectiles at unwary passersby. For me, a snowball is something you eat at the ballpark on a hot, summer afternoon—a ball of ice shavings, smothered in sweet red or purple syrup, and perched in a paper cone. My snowballs are not just bare ice.

Neither are Fred Whipple's. Perhaps the world's foremost authority on comets, Whipple proposed that these spectacular, luminous objects are merely dirty snowballs. But just how dirty are they? Could you eat a comet?

IDEAS ABOUT THE STRUCTURE OF COMETS

Whipple's idea was fairly heretical in 1950, when it was first proffered. In fairness, I should point out that other astronomers had plausible reasons for their own beliefs about comets. It is just not that easy to tell what might be the basic ingredients of these cosmic baubles from telescopic observations. The light by which you see a comet is reflected sunlight, not energy given off by the object itself. Observations of comets made at different times, so that the angular relationship between them, the Sun, and the Earth changed, had convinced Whipple's predecessors that sunlight reflected from comets was strikingly similar to that reflected from dust clouds on the Earth. This similarity led first to speculation and then to a consensus that a comet consisted of loose swarms of separate particles, an orbiting beach of sand grains moving along parallel, closely spaced paths. This idea was logically called the sand-bank hypothesis, and scientific sketches made during the height of its popularity often pictured comets as myriads of closely spaced dots.

These observations indicate that sand particles and dust motes are present in rather large numbers in the comet's head as it approaches the Sun, but they do not really prove that the comet itself is a cloud of particles. After all, advocates of sand-bank comets had only a superficial view of the huge, bright head that makes a comet so distinctive. Whipple argued that the head of a comet has embedded deep within it a tiny *nucleus*, a more compacted object than was envisioned in the sand-bank hypothesis. In his view, once this snowball enters the inner solar system, it becomes completely shrouded in dust and gas. The nucleus is like a shy maiden that undresses only in the dark of interstellar space. If Whipple could not see the nucleus directly, what made him think that it existed?

Astronomers had already learned that comets shed large quantities of matter, both gas and dust, as they approach the Sun. Whipple was convinced that the small amounts of volatile elements—carbon, hydrogen, nitrogen, and others that readily form gaseous molecules—that could be frozen into thin icy coverings on individual dust particles or be trapped in the tiny void spaces within them could not account for this display. Astronomers also had observed jets of matter spraying outward from the sunward side of the heads of some comets, which could not be readily explained by the sand-bank hypothesis. Whipple envisioned the cometary nucleus as a solid aggregation of ice and dust, with ice predominating, and went on to demonstrate how his model could explain several perplexing observations of comet behavior. For example, we might expect loose aggregations of small particles that closely approach the Sun to be vaporized by the Sun's searing heat or disrupted by its immense gravitational field. The observed survival of some Sun-grazing comets seemed to demand a larger, more cohesive object. Deviations in the arrivals of certain comets from their predicted timetables could also be explained by the accelerating or decelerating effects of jets of gas emanating from the daylight sides of rotating snowballs.

Whipple's proposal set off a flurry of activity to try to image a comet's compact nucleus directly and determine its size. The head of a comet is huge, commonly many thousands of kilometers in diameter. How large would a dirty snowball have to be to blossom into such a radiant display? The arrival of a few fortuitous comets, among them Comet Halley, passing directly between the Sun and the Earth seemingly offered opportunities to see the hidden nucleus as a moving dark spot silhouetted against the solar disk. However, Whipple's predecessors had already tried this approach, to no avail. This failure to see the nucleus buttressed their view that Halley and other comets were clouds of particles. It is likely that an object more than 100 kilometers in diameter would have produced a discernable sunspot, so cometary nuclei had to be smaller than this. Astronomers then tried an indirect approach, based on calculating the amount of sunlight that a comet nucleus would reflect. This calculation requires knowledge of two properties of the nucleus, its size and color. Unfortunately, they did not have any information about either property, so such calculations could not possibly produce a unique answer. Assuming for the moment that Whipple's dirty snowballs are filthy and consequently dark and unreflective, their calculated diameters would be just a few kilometers.

Some years later, with the advent of modern radar astronomy, the size of a cometary nucleus could be measured directly by sending a burst of radio waves to probe through the dust halo for a hidden target, which then reflects the signal back to Earth. This technology has demonstrated clearly that Whipple was at least partly right; there is a compact nucleus within each comet head, although its proportion of ice relative to other materials cannot be determined from these data. Radio wave measurements of comet nuclei passing near the Earth have given diameters ranging from several hundred meters to a few kilometers.

Comet nuclei are indeed miraculous objects, kilometer-size chunks that balloon into cosmic features many thousand times their own diameters as they approach the Sun. It's like the tale about the princess and the pea, a tiny speck producing an effect all out of proportion to its size. How does the tiny cometary nucleus manage to create this spectacular display? The fact that a comet becomes brighter within the inner planet region indicates that its luminosity is controlled by nearness to the Sun. If the nucleus really is a snowball, solar heating would be expected to vaporize surface ice, which might erupt in violent jets, carrying with it particles of cometary dust. The *coma*, the bright cloak of gas and dust particles that envelops the nucleus at this point, may grow to be as large as the Earth itself. Some of the tiny particles in the coma would be swept back away from the Sun into one or more graceful tails that stretch for great distances, equal in many cases to the expanses separating the orbits of adjacent planets.

A telescopic view of Halley's Comet during its last appearance illustrates some of these features. Although Halley's nucleus is effectively veiled from astronomical gawkers, a hint of its presence is provided by a computer enhancement of this photograph of the coma, which shows contours representing varying light intensities. Observations at different times indicate that Halley, like other comets, is a quick-change artist, adept at altering its clothing. Its fancy tails are only intermittent features, and there are periods of convulsion when clumps of tail material escape, leaving the coma without its characteristic, trailing decoration. As spectacular as the tails are, they have surprisingly little substance. The collected matter composing a comet's tail, many millions of kilometers long, could probably be contained in the back of a pickup truck.

ORIGIN AND ORBITS OF COMETS

Before we try to answer the question of whether Halley, or any other comet for that matter, is a clean or dirty snowball, let's talk about where it comes from. You wouldn't eat something if you had no idea of where it's been, would you? Our current ideas about the source of comets were focused by the work of Dutch astronomer Jan Oort. Fortuitously, in 1950, the same year that Whipple suggested that comets were dirty snowballs, Oort proposed the existence of a huge

This photograph of Halley's Comet, taken in early 1986, shows a bright ball of glowing gas and dust, with trailing tails of ionized gas and dust. The figure below is the same photograph, now contoured to show different light intensities. (E. Fink, University of Arizona.)

cloud of comets, forming a spherical halo around the Sun extending to a distance of perhaps 50,000 astronomical units.[1] This hypothetical comet repository has now been appropriately dubbed the Oort Cloud. To illustrate the immensity of its distance, the *Voyager* 2 spacecraft, which took 12 years to travel from Earth to Neptune and is now heading out of the solar system, will take another 10,000 years to reach the Oort Cloud. The comets in this region are drifting slowly in roughly circular orbits, taking millions of years to circle the Sun that, at these immense distances, looks like just another star. The Oort Cloud probably contains trillions of comets, but the space they occupy is so vast that it is unlikely that they would ever bump one another. Very little happens there—this is an unexciting place, the very edge of interstellar space. However, the vacuum surrounding the comets is so nearly perfect that it must serve as a relatively clean, cold-storage locker for them.

A vast swarm of icy bodies may reside in the Oort Cloud at the moment, but it seems unlikely that they originally formed there. We know that the giant planets, from Jupiter outward, consist largely of icy materials, somewhat like Whipple's snowballs, and it seems a good bet that comets are leftover debris that somehow escaped being swept up during the formation of these planets. Calculations have demonstrated that gravitational tugs by massive Jupiter and Saturn would have been so violent that unused chunks of matter orbiting within their neighborhoods would have been ejected completely out of the solar system. This same effect has been utilized by NASA flight engineers to modify the orbits of a number of spacecraft, swinging them close to Jupiter so as to accelerate them onward to more out-board planets on their way out of the solar system. Uranus and Neptune are smaller, and the correspondingly milder perturbations they might impart on nearby bodies would be sufficient to eject them only as far as the Oort Cloud. One possible comet that might have escaped this fate is Chiron, an asteroidlike object whose orbit lies between those of Saturn and Uranus. This body exhibits sporadic changes in brightness that might be outbursts of gas and dust, perhaps tantrums of cometary activity.

[1] An astronomical unit, commonly abbreviated as A.U., is 150 million kilometers, the distance from the Earth to the Sun.

If Chiron is a comet that never left its birthplace, the true comets, such as Halley, that we now see in the planetary region are celebrating a homecoming. They formed within the confines of the solar system about 4.5 billion years ago, were banished shortly thereafter to a dark and distant storage site, and are only now reemerging into the light. How could they break free to return to the planetary region? The Oort Cloud is so far away that comets are only loosely bound to the Sun by gravity. Comets in this region are truly on the fringe of interstellar space, perhaps one-third of the way to the next nearest star. A soft gravitational nudge by a passing but still-distant star may be all that is required to reset the orbital path of a particular comet and allow it to come careening into the inner solar system (or off in one of a thousand other directions).

Comets from the Oort Cloud understandably take a long time to make the round trip through the planetary region. On each pass into the inner solar system, there exists the possibility that a comet's orbit may be changed as it encounters the gravitational fields of nearby planets. Comet Halley has an average orbital period of about seventy-six years, but its actual arrival time has varied by as much as two and a half years because of planetary perturbations. Sometimes the gravitational perturbations induced by nearby planets manage to alter a comet's orbit so that it remains constantly within the plane-tary region and never returns to the Oort Cloud. This may occur, for example, as a comet fortuitously passes through Jupiter's orbit at a time when Jupiter happens to be nearby. Jupiter's gravitational field tugs on the comet, deflecting it in such a way that its orbit now traces out a shortened ellipse. Halley's elliptical orbit extends out just a little farther than thirty-five A.U., so it remains completely within the planetary region. (Pluto, the outermost planet, is at thirty-nine A.U.)

THE 1986 RETURN OF COMET HALLEY

So far, anyway, comets haven't been anywhere that should obviously disqualify them as edible material, at least prior to their entry into the planetary region. Before we commit to trying a bite, though, we probably need to take a close look at a comet nucleus, to see if it has suffered any adverse effects from the light show it produces during its close approaches to the Sun. If you are like me, food first has to pass a sight test. It would be highly desirable to have a few cometary ice cubes for careful scrutiny. Unfortunately, that is not yet technologically feasible, although it is being worked on. At the present time we will have to settle for a rather fuzzy view of an entire nucleus, in this case, of Comet Halley.

Halley last returned to the inner solar system in 1986. Like many people, I found the event profoundly disappointing. This comet only comes around once in a lifetime for most of us, and I had harbored hopes of making its sighting a personal experience to be shared with my five-year-old daughter. A memory something like what I wanted for her had already been described by Loren Eiseley,[2] who witnessed Halley's spectacular 1910 traverse through the inner solar system:

> Like hundreds of other little boys of the new century, I was held up in my father's arms under the cottonwoods of a cold and leafless spring to see the hurtling emissary of the void. My father told me something then that is one of my earliest and most cherished memories. "If you live to be an old man," he said carefully, fixing my eyes on the midnight spectacle, "you will see it again. It will come back in seventy-five years. Remember," he whispered in my ear, "I will be gone, but you will see it. All that time it will be traveling in the dark, but somewhere, far out there"—he swept a hand toward the blue horizon of the plains—"it will turn back. It is running glittering through millions of miles." I tightened my hold on my father's neck and stared uncomprehendingly at the heavens. Once more he spoke against my ear and for us two alone. "Remember, all you have to do is be careful and wait. You will be seventy-eight or seventy-nine years old. I think you will live to see it—for me," he whispered a little sadly with the foreknowledge that was part of his nature.

In 1986, an unfortunate turn of events this time around caused the comet to be at its brightest when it was situated on the opposite side of the Sun from the Earth. Even after it emerged from behind the

[2]Reprinted with permission of Charles Scribner's Sons, an imprint of Macmillan Publishing Company, from *The Invisible Pyramid* by Loren Eiseley. Copyright 1970 Loren Eiseley.

solar disk, it was so faint that we never spotted it, and so had to settle for a Comet Halley T-shirt. That is not to say, however, that this last encounter was a complete bust. There were some beautiful telescopic views, but *Giotto* had a front-row seat.

Giotto was a fly-by spacecraft, the property of ESA, the European Space Agency. This small vehicle was named after an Italian painter who incorporated Comet Halley, as it appeared in 1301, into one of his frescoes. The comet was painted in the place of the Star of Bethlehem in an exquisite depiction of the Adoration of the Magi. (The nearest Halley apparition to the birth of Christ was in 12 B.C., which is probably not close enough to make this idea plausible.) Giotto was a pioneer of naturalist painting, and thus his representation of the comet itself is a much more accurate portrait than the stylized sketches of its earlier appearances.

The *Giotto* mission planners were daredevils, sending the spacecraft to within 600 kilometers of the sunlit side of Comet Halley's nucleus. In order to navigate this precisely, one has to know exactly where this tiny moving target is hidden within the coma. Earth-based observations alone can fix the position of the nucleus only to within about 500 kilometers, clearly not precise enough for this feat. ESA, NASA, and their Soviet counterpart[3] combined their considerable talents to pinpoint the orbital parameters of the hurtling nucleus. NASA's radio telescopes recorded transmissions from two Soviet *Vega* spacecraft, which had been launched earlier to first visit Venus and then make passes through the outskirts of Halley's coma at a distance of about 9,000 kilometers from the nucleus. The precise locations of the *Vegas* were computed, and from that the orbital position of the nearby comet was fixed. This information permitted *Giotto* flight engineers to maneuver the spacecraft delicately for a close but not disastrous approach.

From the outset, it was clear that *Giotto*'s dash through Halley's dusty coma at sixty-eight kilometers per second would be risky business, perhaps akin to a kamikaze mission. In order to improve its chances of survival, it was equipped with a meteor bumper, an extra shell of metal placed a few inches outside of the main skin of the

[3]The Soviet Union had no central organization for space activities comparable to NASA or ESA. *Vega* spacecraft were operated by the Space Research Institute (acronym IKI in the Cyrillic alphabet).

spacecraft. This bumper might protect against tiny particles, but no one knew whether to expect just some pesky sand-blasting or a fatal collision with a football-size dirt clod. *Giotto* did encounter a lot of small dust particles, finer than cigarette smoke, but was spared from larger impacts.

What *Giotto* glimpsed as it raced through this haze was the Halley nucleus itself. Its camera revealed a sixteen-by-eight-by-eight kilometer potato, its sunward side violently spewing away each second about twenty tons of gas and between three and ten tons of dust. Its surface is as dark as black velvet, reflecting only a few percent of the sunlight falling on it. In places this opaque crust appears to have cracked open, and out of these fissures erupt thick jets of dust. Its surface is rough, as judged from the occurrence of a 400 meter–high hill jutting up into the sunlight and a depression that probably represents the site of some previous outburst. It rotates with a period of about fifty-four hours, but its complicated shape also causes it to wobble.

All this is not exactly encouraging from a culinary perspective. But before we get too disheartened, let's see what we can learn about the actual composition of the comet.

The Halley nucleus weighs on the order of 100 billion tons, which is somewhat greater than estimates made before the nucleus had been imaged. From a comparison of this mass with the amount of matter it loses during a typical pass through the inner solar system, we can safely infer that Halley is in no danger of running out of material in the next few thousand years. By combining photographs of the nucleus taken by *Giotto* and by the more distant *Vegas*, it has been possible to construct a three-dimensional likeness, which has a volume of about 500 cubic kilometers. Its estimated density, determined by dividing the mass by the volume, is then just a few tenths of a gram per cubic centimeter. For comparison, water has a density of one gram per cubic centimeter and ice is only slightly less dense. Comet Halley must be very porous or have some other lighter constituents to account for its low density.

We know that a substantial fraction of the nucleus is made of ice, frozen water presumably like that we encounter every day. It did not start out that way, however. Freezing water vapor at very low temperatures, such as those in the neighborhood of Uranus and Nep-

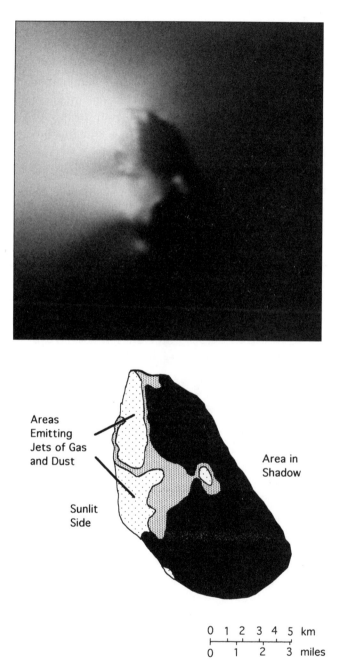

The ghostly image of Comet Halley, recorded by the *Giotto* spacecraft as it dashed through the coma, reveals a potato-shaped nucleus spewing gas and dust from its sunlit side. The accompanying sketch identifies some important features visible in this blurry photograph. *(European Space Agency.)*

tune, produces what is known as amorphous ice. This is a solid in which the constituent water molecules are not arranged three-dimensionally in any fixed geometrical array, as they are in ice crystals. Amorphous ice is not a particularly stable configuration, and upon slight warming it transforms to the more familiar crystalline ice. Comet nuclei arriving in the planetary region for the first time often have spectacular comas, probably enhanced in part by the energy given off when their amorphous ice crystallizes in the warmth of the Sun.

Giotto and the *Vegas* carried aboard them sensitive instruments to identify gaseous molecules in Halley's coma. What they found, confirming conclusions that had already been made from telescopic observations, was a veritable stew of molecular fragments called free radicals. Formed when molecules are split apart by solar radiation, these torn pieces of molecules are highly reactive, tending to recombine with other free radicals or break apart further into smaller, more stable molecules. The free radicals can be detected within the tenuous comet coma, however, because they are far enough apart that they have appreciably longer lifetimes than they would on Earth. The spacecraft analysis of Halley's coma was dominated by molecular fragments formed from water, as Whipple predicted. Some 80 percent of the volatile fragments were formed from water ice, but that was just the tip of the iceberg. Other free radicals indicated that about 10 percent of Halley's ices are carbon monoxide. Piecing together free radicals to discover what the parent ices were is tricky business, but such molecular puzzles suggest that frozen carbon dioxide, ammonia, methane, and hydrogen cyanide are probably present as well. Some of these may be physically trapped inside cages of frozen water molecules, producing a structure known to chemists as a clathrate.

Somewhat surprisingly, not all of the volatile elements in Halley's coma were in gaseous form. A major discovery was the existence of tiny, solid CHON particles. CHON is an acronym for the chemical symbols of the elements of which these particles are composed: carbon, hydrogen, oxygen, and nitrogen. These particles are apparently organic compounds. The term organic simply means that these are compounds based on carbon, not that they are necessarily biological in origin. When combined with other elements, organic molecules commonly take delicate chainlike or branching forms. Just as free radicals in the coma can be used to tell us of the presence of ices in the nucleus, CHON particles signal the presence of organic compounds in it. These might appear as black soot or dark reddish gooey sludge, probably mixed intimately with the ices.

Dust detectors on the *Giotto* and *Vega* spacecraft found large quantities of very fine particles, generally smaller than a few micrometers (a micrometer is one-millionth of a meter) in size. Chemically, these motes consist of many of the same elements that comprise terrestrial rocks: iron, magnesium, silicon, oxygen, and others as well. Because of their small size, determining a representative composition for the dust is challenging, and guessing at the minerals that comprise them is even more difficult. Nevertheless, it is clear that the Halley nucleus contains abundant rocky dust. It is this material, along with the organic compounds, that apparently has been concentrated as a black veneer on the surface of the nucleus, an unsightly residue left behind as the ices volatilize in the presence of solar radiation and escape to space. Some scientists have suggested that dust is the dominant material inside comet nuclei, making them more like frozen mudballs than dirty snowballs.

INTERPLANETARY DUST

However wonderful the data gleaned from the *Giotto* and *Vega* spacecraft were, we still have not properly characterized the material that makes the Halley nucleus dirty. But there may be another effective, though slightly less adventurous, way to get a handle on the nature of cometary dust. We can wait for it to come to us or, more accurately, wait for the Earth to come to it. A comet is like a trashy tourist, leaving in its wake a glistening trail of grime. These particles travel around the Sun at different velocities, so that over time they spread out to form a continuous stream that traces out the orbit of the comet. If this stream happens to intersect the orbit of the Earth, a portion of this dusty residue will be left in our path, just waiting to be swept up by the Earth on its next pass.

The larger dust particles enter the atmosphere at high speed, so friction with the air rapidly burns them into cinders. When the Earth encounters many such particles in a short time, a meteor shower is produced. Very small particles a few micrometers in size can radiate their frictional heat more efficiently so, rather than burn up, they simply slow down in the upper atmosphere, gradually sifting down to the Earth's surface. These particles are all around us, but they are hard to recognize and even harder to collect because their concentrations on potential collecting surfaces are swamped by atmospheric pollutants. There are a few places on our planet, such as Antarctica, that are clean enough for particle collections, but where most of us live that is impossible.

A more effective means of collection is to grab these particles in the upper atmosphere, before they filter down into our industrialized

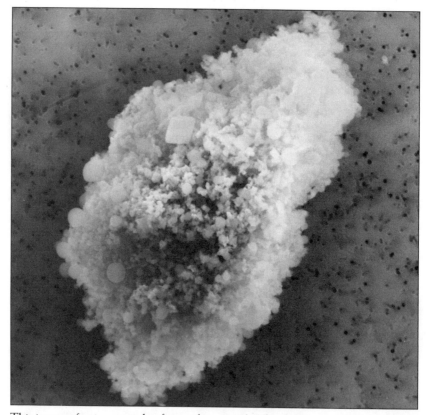

This image of a tiny particle of interplanetary dust has been magnified thousands of times by an electron microscope. The particle was trapped in the upper atmosphere during a U-2 flight. This dust mote might have once been part of a comet nucleus. (*NASA Johnson Space Center.*)

air. This is accomplished by a U-2 aircraft, a high-flying machine originally designed for surveillance (actually spying) but now rededicated to scientific research. Attached to a pylon mounted on its wing is a collector plate smeared with silicone grease. The collector is not opened to the airstream until the U-2 climbs into the upper atmosphere, and it is covered again before descent. As it flies, the collector plate catches interplanetary dust particles (IDPs in abbreviated form, or sometimes "Brownlee particles" in reference to the scientist who pioneered this procedure), immobilizing them on its sticky surface. The day's haul of IDPs is revealed back on Earth only when the plate is examined under a microscope.

Magnified thousands of times, many IDPs are revealed to be delicate, fluffy aggregates of even smaller grains. Among the loosely packed smaller grains are minerals familiar from terrestrial geological experience—various kinds of silicates, sulfides, and oxides. Also present are tiny pockets of organic material, possibly CHON as it would appear close up. We do not know for sure if IDPs are cometary dust, but it's a good bet that some of them are. The IDPs that are not cometary have been ground off the surfaces of rocky bodies in the asteroid belt and have gradually made their way into the inner solar system.

BUT IS IT EDIBLE?

So let's return to our original question—could you eat a comet? From what we have just learned, it would not appear to be a wise idea. Periodically returning comets with predictable orbits would probably be the only ones that might offer any hope of direct sampling. The appearance of the Halley nucleus suggests that the exteriors of such objects have been altered by repeated close passages near the Sun. To find some cleaner ice, you would have to excavate down through a crust of compacted dust and organic material. Besides being a dirty job, this might prove to be hazardous. As soon as you had plumbed through to ice, sunlight might cause it to start vaporizing instantly, creating a violent jet of gas and stinging particles. If you could somehow sample the ice safely, you would find that it was not just water ice, and carbon monoxide ice cream might be a little exotic for most palates. The ice might also have an unappealing reddish color and possibly an unpleasant flavor, both the result of organic compounds intimately mixed with it. If the ice is very porous, as seems likely in the interior of the nucleus, then it would melt too quickly to make useful ice cubes anyway. If hydrogen cyanide is a component of the ice, it might have a pleasant almond smell but would be poisonous. The fine-grained particles of dust in the cometary nucleus also would make it a less-than-desirable foodstuff; this grit-and-ice mixture would be a poor substitute for the ice with which we are familiar, and eating it might require a good dental insurance plan. I suppose you could eat a few carefully selected and processed bites of a comet, but I think I'll pass.

Some Suggestions for Further Reading

Balsiger, H., Fechtig, H., and Geiss, J. 1989. "A Close Look at Halley's Comet." *Scientific American*, vol. 259, no. 3, pp. 96–103. Summarizes the results of the *Giotto* mission to Comet Halley in early 1986.

Littmann, M., and Yeomans, D. K. 1985. *Comet Halley: Once in a Lifetime*. Washington, D.C.: American Chemical Society. This is an excellent and highly readable reference on what was known about this comet before the *Giotto* encounter.

Moore, P., and Mason, J. 1984. *The Return of Halley's Comet*. New York: Norton and Company. Another of the flurry of books predating Halley's 1986 return, a well-illustrated introduction to comets.

Olson, R. J. M. 1985. *Fire and Ice: A History of Comets in Art*. New York: Walker and Company. A beautifully illustrated monograph on artists' depictions of comets.

Sagan, C., and Druyan, A. 1985. *Comet*. New York: Random House. Nobody else writes like this. This is an engrossing and entertaining examination of everything you ever wanted to know about comets.

Whipple, F. L. 1985. *The Mystery of Comets*. Washington, D.C.: Smithsonian Institution Press. A highly readable and authoritative treatment of the subject by the dean of American comet researchers.

Wilkening, L. L., ed. 1982. *Comets*. Tucson: University of Arizona Press. A reference for those interested in a more technical treatment, this book contains twenty-eight papers by leading comet researchers.

Fire and Ice

Active Volcanism on
Io and Triton

Sometimes, a choice we initially thought to be black or white instead may turn out to be some shade of gray. Opposites are not always contrary in every way, as illustrated by two bizarre and rather inhospitable moons of the giant planets Jupiter and Neptune. In appearance these moons, Io and Triton, are about as different from each other as any two objects could be. Io is fire, a hellish place painted lurid red and ocher; Triton is ice, the coldest object in the solar system, so far as we know. Nevertheless, Io and Triton perversely share a common link: They are the only satellites understood to be volcanically active at the present time.

HEAT SOURCES

Living on a world with active volcanoes may have lulled you into thinking that this phenomenon is commonplace, but a tour around the solar system demonstrates that it is not. From a planetary perspective, volcanism is much more than just a geologic curiosity. Erupting lavas signify the presence of internal heat, a perpetual furnace potent enough to melt the planet's rocky substance. The fuel for the Earth's heat engine is the very slow decay of radioactive isotopes, unstable atoms that give off small quantities of heat as they break apart. The natural radioactivity of potassium and uranium atoms within a marble-size piece of granite generates only enough warmth in a million years to brew a cup of coffee, yet the temperature of that same rock can climb to 1,000 degrees centigrade or more over the lifetime of the Earth, provided that the heat cannot escape. This can be accomplished by placing the marble deep inside the planet. Rocks are notoriously poor conductors of heat, so a large, rocky body is an excellent insulator that traps the heat slowly released during radioactive decay in the planet's interior. The more rock, the better the insulation, so the Earth's large size accounts for its periodic volcanic eruptions.

Radioactive decay is also the primary source of heat in other large planets, but smaller bodies without sufficient rocky insulation cannot trap heat for long periods of time. The Earth's Moon is a prime example of the importance of planet size in maintaining internal heat. The Moon has been volcanically dormant for several billion years, because it is too small to have sequestered its radioactive heat until the present day. Surprisingly, Io and Triton are each roughly the same size as the Moon, yet both are currently volcanically active. How is this possible?

An intriguing solution, at least for Io, was discovered by a trio of planetary scientists in California. Stanton Peale of the University of California at Santa Barbara, working with NASA researchers Patrick Cassen and Ray Reynolds, studied the gravitational forces acting on this moon. Io is the innermost of Jupiter's major satellites, and its proximity to the giant planet suggests an explanation for its volcanism. The combination of strong gravitational pulls by Jupiter and the nearby moon Europa raises a tidal bulge on Io, much like the ocean tides produced on Earth by the tugging effects of the Sun and Moon. But there is no ocean on Io, so the bulge occurs on the satellite's surface. Additional tugs by Io's more outboard sister satellites cause it to oscillate continuously in its orbit, first inward toward Jupiter and then outward again. This oscillation flexes Io's rocky bulge, and an enormous amount of frictional heat is transmitted into the satellite's interior. Peale and his coworkers reasoned that this tidal heating would cause widespread melting, which should be revealed on the surface in the form of volcanic eruptions.

The bold prediction that Io would be volcanically active was published in 1979, appearing in print just three days before *Voyager 1* arrived at Jupiter and provided the first close look at its moons. This kind of scientific bravado is not uncommon in planetary research, though, and is even encouraged to some degree. On the inside cover of each issue of *Icarus*, a prominent journal of planetary science, is the following quotation:

> In ancient days two aviators procured to themselves wings. Daedalus flew safely through the middle air and was duly honoured on his landing. Icarus soared upwards to the Sun until the wax melted which bound his wings and his flight ended in fiasco. . . . The classical authorities tell us, of course, that he was only "doing a stunt"; but I prefer to think of him as the man who brought to light a serious constructional defect in the flying-machines of his day. So, too, in science. Cautious Daedalus will apply his theories where he feels confident they will safely go; but by his excess of caution their hidden weaknesses remain undiscovered. Icarus will strain his theories to the breaking-point till the weak joints gape. For the mere adventure? Perhaps partly, that is human nature. But if he is destined not yet to reach the Sun and solve finally the riddle of its construction, we may at least hope to learn from his journey some hints to build a better machine.[1]

The adoption of Icarus' name for this journal is meant as subtle encouragement to planetary scientists, whose stock and trade are often inferences drawn of necessity from meager data.

In any case, Peale and company didn't have to wait long to find out if their speculations were correct. The first pictures taken by *Voyager 1* as it rushed past Io showed a number of small, dark spots. On first impression, these appeared to be similar to meteor impact craters, which must have given our brave trio some pause. They were anticipating a smooth volcanic plain. However, higher-resolution im-

[1]The quotation was adapted from Sir Arthur Eddington, *Stars and Atoms*, Oxford University Press (1927). By permission of Oxford University Press.

ages quickly dispelled the notion that these spots were due to impacts. In reality, the spots were calderas, collapsed craters produced when lavas were emptied from underground chambers. Further searching showed that this volcanic surface contained no impact craters at all! Any impact structures were apparently completely buried beneath voluminous lava flows, suggesting that volcanism must have been very recent.

ERUPTIONS ON IO

Voyager 1 hurtled through the Jupiter system in thirty hours, but the breathtaking images it sent back to Earth kept scientists awake for several days running at mission control. Hours after the last of them had finally stumbled exhaustedly into bed, Linda Morabito made her own startling discovery. Morabito was an engineer assigned to the *Voyager* navigation team's graveyard shift, and one of her responsibilities was to obtain a precise fix on the spacecraft's location. In examining some photographs of Io that she had overexposed to bring out faint stars in the background, she noticed a peculiar umbrellalike form on the satellite's southern limb. At first she thought this ghostlike visage might be some artifact, possibly the image of another satellite partly hidden behind Io. The position of the spacecraft, however, precluded any of the other moons appearing in line with Io. What Morabito had really discovered was a huge volcanic plume arching hundreds of kilometers above the satellite's surface. Now there was no disputing the fact that Io was volcanically active.

Once alerted to the occurrence of this erupting plume, scientists wasted no time in initiating a systematic search for others. They quickly discovered eight more active volcanoes spewing forth clouds of powdery dust and gas. Io wasn't just volcanically active; overnight, it was recognized as the most phenomenally active object in the solar system. Of course, the nine plumes had not just switched on during the night of Morabito's discovery. They had been there all along, but the mists of fine particles that composed them were not visible until enhanced by digital processing. All but one of the plumes found by *Voyager 1* were still active when *Voyager 2* arrived on the scene four months later. The huge plume discovered by Morabito, which had

48

been named Pele for the Hawaiian god of volcanoes, had shut down. A new volcano had apparently deposited a blanket of material and then extinguished itself during the interval between the *Voyagers'* visits, a spurt that was over almost before it had bugun. Evidently, some of these fountains have lifetimes of only a few months.

Typical plumes are shaped like opened umbrellas, or sometimes broken umbrellas, as not all of them are symmetrical. These fountains have been likened to geysers on the Earth, but that hardly does them justice. If Old Faithful could be magically transplanted from Yellowstone National Park to Io, with its much lower gravity and atmospheric pressure than on Earth, its steam would skyrocket to a height of 38 kilometers, twice as high as the maximum cruising altitude of commercial airliners. Even so, this display would be trivial compared to the largest Ionian plumes, which reach heights of 300 kilometers and spread out laterally to widths of 1,200 kilometers. Each second, a typical Ionian volcano spews out 1,000 tons of mate-

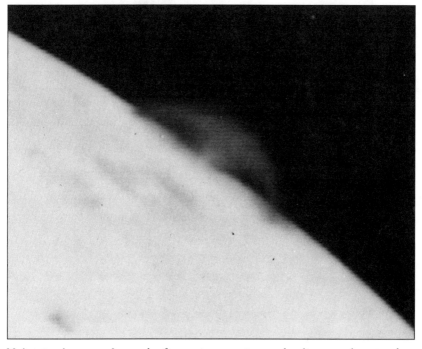

Volcanic plumes on Io are the first active eruptions to be discovered on another solar system body. This plume was photographed twice by *Voyager 1* over a two-hour period. Its distinctive, umbrellalike form extends more than 100 kilometers above the surface. (*Jet Propulsion Laboratory.*)

rial. These ejected particles follow ballistic trajectories, like tiny artillery shells, which ultimately curve back to the surface to form the distinctive parasol shape.

Tracing an Ionian plume back down to its source on the ground, we see a dark, gaping pit circumscribed by rings of material rained down from above. The region immediately below the Pele plume, the largest of those observed by *Voyager*, is easily the most dramatic surface feature on Io. Light and dark markings delineate a gigantic heart with a dark, central core. But I can assure you that this is no valentine. The outermost black ring coincides with the diameter of the plume itself, so it may represent the most recent deposit of falling material. The multiple rings probably record progressive eruptive stages in the volcano's evolution, something like growth rings in a

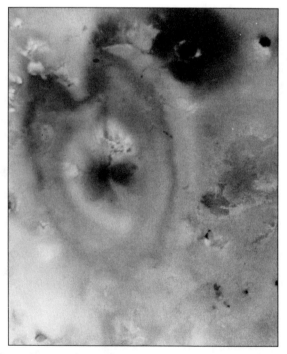

This heart-shape deposit of pyroclastic powder rained down on Io's surface from the Pele plume, the largest observed on Io. The outer dark ring is 1,000 kilometers across in its longest dimension, about the same as the width of the plume itself, so this may represent the most recently fallen material. In the center is a dark pit, from which the plume erupted. *(Jet Propulsion Laboratory.)*

tree, although adjacent rings need not have formed sequentially. The dark pit at the center of the heart is an elongated depression, a caldera twenty-four kilometers across at its widest point, which is surrounded by dark volcanic deposits extending outward for several hundred kilometers.

The volcanic materials around this and other similar pits are *pyroclastic* (from the Greek words *pyros* meaning "fire" and *klastos* meaning "broken"), a term used by geologists to describe fragments that are explosively ejected from volcanoes on the Earth. Pyroclastic eruptions are particularly violent, and most volcanic disasters are associated with this kind of volcanic behavior. The Roman cities of Herculaneum and Pompeii were entombed by hot pyroclastic ash from Mount Vesuvius, and, much more recently, settling pyroclastic powder from Mount St. Helens turned several states in the Pacific Northwest of

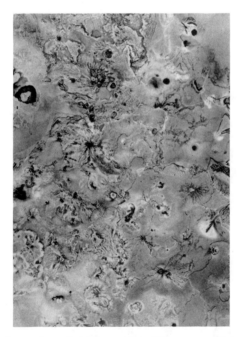

This photograph shows about 2 million square kilometers of Io's surface. Visible volcanic features include cone-shape mountains with pits on their summits, collapsed calderas, and sinuous lava flows. The smooth background on which the volcanoes and flows are perched is probably a blanket of pyroclastic material. The absence of impact craters indicates that this surface is very young. (*U.S. Geological Survey.*)

the United States an ashen color. During such eruptions, particles of all sizes are lofted high into the air, some even reaching into the stratosphere, and are aerodynamically sorted by size during flight. The finest material generally travels the greatest distance before it settles to the ground.

Other manifestations of Io's volcanic personality include cone-shape edifices similar to the largest volcanoes on Earth. Elsewhere, sinuous rivers of lava have poured from black wounds onto the Ionian surface. Similar-shape lava flows occur on the Earth, but Io's flows are as much as 200 kilometers in length, more than ten times as long as comparable flows from terrestrial volcanoes.

Io's surface, then, is being continually repaved by lava flows and blanketed by deposits of pyroclastic material. Lava flows, at least ancient ones, are a common occurrence on many planets and some satellites, but explosive plumes are not and require some further explanation. Violent pyroclastic eruptions on the Earth are powered by expanding gases, mostly water vapor but others as well. These gases are dissolved in magmas deep in the Earth's interior, where pressure from the weight of overlying rocks increases their solubility. As the magmas ascend to the surface and the pressure on them diminishes, the gases form bubbles, much like the frothy effervescence of soda pop when its confining pressure is released by opening the bottle. The rapid formation and expansion of the gas bubbles is like a gigantic burp that forcefully ejects bits of congealed lava and may even pulverize a portion of the overlying volcano itself. Pyroclastic eruptions also occur when ascending magmas intrude directly into the water table, exploding through the surface as large quantities of underground water are vaporized instantly. Could one of these mechanisms possibly explain plume formation on Io? There is no water on this satellite to serve as a propellant, but there is plenty of something else just as good: sulfur.

Io's lavas come in a bewildering variety of colors, ranging from black to greenish yellow and brownish orange. Various forms of sulfur can take on all of these hues. Sulfur is an unusual element in that its atoms can bond together in different structural forms, called allotropes, and various sulfur allotropes produce a riot of color. Sulfur melts when heated to 115 degrees centigrade, just a little hotter than the boiling point of water, and the higher its temperature, the deeper its color becomes. It retains the color of the hot liquid when rapidly cooled into a solid. Io's garish colors have prompted some scientists to propose that the floors of the calderas are pools of encrusted liquid sulfur, which pours out periodically to form the sinuous lava flows. Another view is that the lavas are silicate magmas like those on the Earth, but they are coated, like almost everything on Io, with pyroclastic sulfur powder. Both of these ideas might be correct, depending on where you look on Io, and some flows could even be sulfur-silicate mixtures.

Sulfur, in the form of smelly sulfur dioxide, also occurs in Io's tenuous atmosphere. This noxious gas is a familiar culprit in the industrial pollution of the Earth's atmosphere. My own experience with sulfur dioxide harkens back to the late 1950s, when I got my first chemistry set. The synthesis of even trace amounts of this pungent gas in my mother's kitchen could always be counted upon to get an immediate reaction, usually my banishment from the house. The pressure of Io's sulfur dioxide atmosphere is minuscule, perhaps one ten-millionth of the Earth's atmospheric pressure at sea level, but the existence of this compound is nonetheless significant. Io's atmospheric gases are continuously lost to space and so must be replenished constantly by volcanic activity.

Sulfurous compounds are clearly abundant on Io's surface and in its atmosphere, so it seems likely that they also occur underground. Below ground level there must be reservoirs for the erupting lavas and plumes, vast subterranean lakes of molten sulfur and sulfur dioxide. Io is certainly not all sulfur, though. The density of this satellite is about the same as that of the Moon, so it must be composed substantially of silicate rock. Tidal heat in the rocky bowels of Io presumably causes melting of this rock to produce silicate magmas, perhaps similar in composition to those on the Earth. Once formed, these silicate magmas would ascend toward the surface, there encountering the molten, sulfurous pools. The consequence is probably analogous to the Earthly situation when magma invades the water table. Rapid heating causes sulfur dioxide to vaporize, and its expansion propels the gas and molten material through vents to the surface. On eruption, some fraction of the sulfur gas condenses into a spray of solid particles, a fine sulfur snow that then falls back to the surface in plumes. So much sulfur is pumped out of Io's interior by this process, however, that some remains as a gas, creating the thin, temporary atmosphere. This gaseous envelope gradually escapes to space, and, over time, it has thoroughly contaminated Io's orbit, producing a doughnut-shape ring of sulfur atoms surrounding Jupiter.

By now you may have concluded, quite correctly, that Io is no vacation spot. In reality, it is an inhospitable and dangerous place, but it may not always have been so. Initially Io was, in all probability, similar in composition to the other satellites of Jupiter, which are composed partly of rock and partly of ices. But Io's runaway temperature, a fever driven by tidal forces, has altered its surface composition profoundly. Over time, it lost its nitrogen, water, and carbon dioxide, which account for most of the ices in the other satellites, as they were vaporized or volcanically ejected to high altitudes where Io's weak gravity could not hold them. Once its original inventory of icy matter had been driven off, sulfur became the next substance to be flushed from the satellite's interior. Sulfur has a lower melting point than the other constituents of silicate rocks, so it was segregated as a liquid that ascended toward the surface. Persistent volcanic activity has now concentrated this element in Io's crust and on its surface. I think it is particularly fitting that this hellishly volcanic moon is, at least superficially, a sulfur world. An antiquated term for sulfur is "brimstone," and the biblical hell is described as a place of fire and brimstone.

AN ICEBOX WORLD

Compared to Io, Triton appears deceptively benign. *Voyager 2* snapped a few pictures of this, the largest moon of Neptune, as the spacecraft rushed past in 1989 on its way out of the solar system. Triton is comparable in size to Io, but there the similarity ends.

Triton's surface temperature hovers around thirty-eight degrees above absolute zero (minus 237 degrees centigrade), the most frigid body so far encountered in the solar system. A veneer of methane and nitrogen ices, along with a substrate of water ice, glazes Triton's surface. Such exotic ices require extremely cold temperatures, and so are not stable on warmer satellites and planets. The common form of methane on the Earth is as a gas, sometimes seeping from marshy swamps. Nitrogen gas comprises the bulk of the Earth's atmosphere, and if our planet were somehow repositioned into Triton's orbit, this gas would condense to form a global shell of nitrogen ice some fifteen meters thick. Liquid nitrogen is produced artificially on Earth for use as laboratory coolant, but making nitrogen ice is something of a technological feat. At one time it was suggested that Triton might be awash with oceans of liquid nitrogen, as would befit an object named for the son of the sea god Neptune. The stark reality, though, is that it is just too cold.

Another of Triton's oddities is the way it orbits Neptune. Its orbit is retrograde, that is, it revolves about Neptune in the opposite direction to the planet's rotation. Triton is the only moon in the entire solar system that does this. The plane of Triton's orbit is also inclined at an unusually steep twenty degrees from Neptune's equator. Many scientists believe that this strange orbit indicates that Triton is a captured satellite, one formed elsewhere and gravitationally snagged during a close approach to Neptune in the distant past. However it came to be a moon of Neptune, it now occupies an unstable orbit. Triton's gravity raises a tidal bulge on Neptune, just as Jupiter does on Io. Because Neptune rotates in the opposite direction to Triton's revolution, the bulge tends to lag slightly behind rather than to be located directly opposite to Triton. The net effect is that the satellite is slowed, slightly but perceptably, with each revolution. This decrease in velocity translates into a gradual spiraling down of the satellite toward Neptune. Luckily for Triton, and for Neptune as well, the effect is too small to present a danger for billions of years.

On top of all this, Triton is a tilted world, unlike most planets and satellites. Its spin axis is tipped almost twenty-nine degrees toward the Sun. This tilt and the inclination of its orbit act together to cause Triton's poles to point alternately toward and away from the Sun during a complex cycle of seasons that lasts nearly 600 years. First one pole and then the other basks in sunlight or, conversely, hides in the dark.

These seasonal oscillations create a large polar cap of nitrogen ice on the shadowed side. This ice evaporates when summer arrives, and the vapor recondenses at the opposite pole. The southern polar cap is experiencing spring at the moment, and summer will not arrive in the southern hemisphere until the year 2000. This polar cap has

The south pole of Triton, shown in the lower part of this *Voyager* 2 composite photograph, is covered with a cap of nitrogen ice. An unusual terrain above the polar cap has the dimpled and lined appearance of a cantaloupe rind. Smooth areas at the equator near the shadowed edge are frozen lakes. (*Jet Propulsion Laboratory.*)

already begun a retreat, and the clouds and haze that *Voyager* 2 observed in Triton's atmosphere may be part of this seasonal polar pilgrimage.

The polar cap, at least for the time being, overlies a strangely textured region that offers a pretty fair imitation of the rind of a cantaloupe. The cantaloupe terrain has dimples that are crosscut by ridges, which formerly were cracks where the surface was pulled apart and then filled with oozing volcanic material that welled up from below. This material might have been a solution of water and ammonia, something like household cleaning fluid but with a greater concentration of ammonia.

The cantaloupe terrain is dotted in places by smooth plains that appear to be huge frozen lakes. Water-ammonia lavas, made thick and slushy by the bitterly cold surface temperature, once may have flooded large regions of Triton's surface. The surfaces of the earliest ponds solidified to form a crust, which then foundered and refroze. Repeated freezing of a surface layer and collapse to expose the underlying liquid have produced a distinctive pattern composed of thin ice terraces stacked one on another.

These frozen lakes and icy fracture fillings suggest that some heat source must have operated in Triton's past. As with Io, tidal forces may have been the cause. Neptune's capture of Triton and the subsequent gravitational taming of this wild moon into a circular orbit could have melted Triton's ices and produced lavas composed of liquid water and methane. After the regularization of Triton's orbit, these tidal forces subsided, and icy volcanic activity ground to a halt.

ERUPTIONS ON TRITON

The stabilization of Triton's orbit should have caused the moon to cool and become dormant, yet *Voyager* 2 found otherwise. Specially processed photographs of the southern ice cap reveal a few dark streaks, some as long as 100 kilometers, and all oriented in the same direction. The streaks are dark only in a relative sense. Under normal viewing conditions they are actually as reflective as snow, but the background of nitrogen ice is even brighter. Each streak begins as a sharp point and widens into a diffuse cloud.

Satellite photographs of North America during the eruption of Mount St. Helens showed a similar cloud of dust being driven eastward by the prevailing winds. The ash cloud began at a sharp point, corresponding to the location of the volcano's summit, and became more diffuse as it was carried into the atmosphere above neighboring states. Beneath this black cloud, western Washington and Idaho were inundated with volcanic ash that choked people, animals, and automobiles.

The dark streaks on Triton also are apparently the results of volcanic or geyserlike eruptions. Dark, roughly circular vents can be seen in the ice cap at the apex of some of these volcanic streaks. Eruptive clouds that gushed upward from these vents were dragged downwind by gentle atmospheric circulation until they settled on the surface. Like Gretel's brother Hansel, who left a trail of bread crumbs to

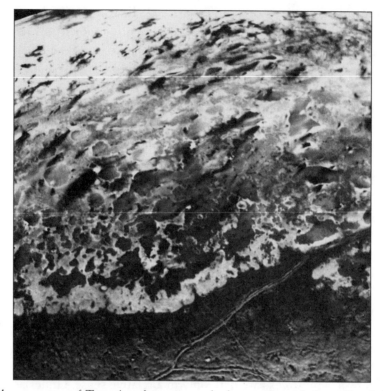

A close-up view of Triton's polar cap reveals the presence of about fifty dark streaks. Most of these streaks are surface deposits of material ejected from small volcanoes at their northeastern ends. This material erupted to a height of eight kilometers, was then carried away by atmospheric circulation, and finally dropped onto the surface. Two of these streaks are materials carried aloft by plumes from presently erupting volcanoes. *(Jet Propulsion Laboratory.)*

follow home, each Triton plume marked its path by dropping fine powder on the bright surface below.

Because these volcanoes were constructed on the temporary polar ice cap, it was clear from their first discovery that they must be very young. Later detailed analysis of photographs revealed that several of them were actively erupting at the time of the *Voyager* encounter. The three-dimensional character of their plumes can be discerned by using a *stereoscope*. This instrument shows a simultaneous view of two photographs of the same place taken from slightly different vantage points, so that vertical features are revealed. Stereoscopic analysis of *Voyager* photographs showed two dark stems rising vertically to heights of about eight kilometers. Abruptly at that altitude, the tops of these stems were swept into long trails.

What could account for this volcanic display on the polar cap of such a bitterly cold world? Understanding the physical behavior of nitrogen, which forms the bulk of the polar ice cap, may be helpful at this point. With only a slight increase in temperature, nitrogen ice will sublime to produce nitrogen gas. As nitrogen vaporizes, it expands and exerts pressure on its surroundings. In fact, a temperature increase of only ten degrees centigrade above Triton's very low surface temperature results in a hundredfold increase in gas pressure. Nitrogen gas, formed and confined inside the ice cap so that its pressure increased, could rupture the overlying ice and produce the observed volcanic plumes. But the idea that the plumes are mostly nitrogen begs the question. How was the nitrogen ice heated?

Present-day volcanism on Triton appears only on the polar ice cap near the latitude that receives maximum illumination from the Sun, and therein may lie an explanation. The *Voyager* imaging science team, led by Bradford Smith of the University of Arizona, proposed a novel heating mechanism for this satellite. In a 1989 paper in *Science*, Smith and his sixty-three coauthors (which must be some kind of record!) suggested that violent vaporization of nitrogen in Triton's polar cap was caused by sunlight. What is needed to make this explanation work is a greenhouse. On Earth, the glass windowpanes that cover a greenhouse let in sunlight that, once inside, is transformed to heat energy and cannot escape. The greenhouse envisioned for Triton employs a windowpane of translucent nitrogen and, below that, nitrogen and methane ice darkened by organic material that would absorb the solar energy. This icy configuration might allow the pressure of nitrogen gas to increase in the warmed lower ice. Rupturing of the icy windowpane would result in violent escape of the trapped vapor, launching a plume of gas and ice particles. This plume might tear off dark particles from its nozzle and carry them aloft, which would subsequently settle to produce the observed dark trail.

Another possibility is that the volcanic outbursts might be powered by heat released when frozen nitrogen changes from one crystalline state to another. Several forms of nitrogen ice are thought to exist on Triton, lending support to this idea. While the volcanoes on Triton are not so spectacular in appearance as are those on Io, the proposed mechanisms that drive them are equally remarkable.

A VOLCANIC DUO

There is no getting around the fact that Io and Triton are different. Io looks, at least superficially, like some gargantuan pizza, smothered in melted cheese and colorfully decorated with pepperoni and ripe olives. Triton is a frost-bitten cantaloupe. Each is unique in its own way, but they also share a profoundly important characteristic. Unlike all other satellites in the solar system, and most planets for that matter, they host active volcanoes. Neither of these unusual moons is big enough to have retained much heat from the slow decay of radioactive isotopes, which is the conventional explanation for active volcanism. We infer, from knowledge of our own Moon and other similar-size bodies, that both Io and Triton should by all rights be geologically dead, but they are not. Io's present-day volcanic activity is due to the gravitational forces exerted on it by Jupiter and its sister satellites, interactions that have produced a sulfurous inferno. An early era of volcanism on Triton also may have been powered by Neptune's tidal forces, but its current activity results from gentle warming of some bizarre ices under the faint glare of a faraway Sun or from sudden rearrangements of the ice's crystalline form.

Among the many moons visited by spacecraft, we have discovered only two, Io and Triton, that are volcanically active. Perhaps fire and ice are not really so different after all.

Some Suggestions for Further Reading

Hamblin, W. K., and Christiansen, E. H. 1990. *Exploring the Planets*. New York: Macmillian. A superb introductory text, with up-to-date sections on both Io and Triton.

Morrison, D., ed. 1982. *Satellites of Jupiter*. Tucson: University of Arizona Press. Lots of mostly technical information about the major satellites of Jupiter. Chapters by G. G. Schaber on the geology of Io and S. W. Kieffer on Ionian plumes are particularly interesting.

Smith, B. A., et al. 1979. "The Galilean Satellites and Jupiter: Voyager

2 Imaging Science Results." *Science*, vol. 206, pp. 927–950. A technical paper summarizing the imaging science team's discoveries from Io and other satellites of Jupiter.

Smith, B. A., et al. 1989. "Voyager 2 at Neptune: Imaging Science Results." *Science*, vol. 246, pp. 1422–1449. An authoritative overview of *Voyager* 2's encounter with Neptune and its satellites.

Soderblom, L. A. 1980. "The Galilean Moons of Jupiter." *Scientific American*, vol. 242, no. 1, pp. 88–100. This paper provides a good, nontechnical description of the *Voyager* encounters with Io and the other satellites of Jupiter.

Soderblom, L. A., et al. 1990. "Triton's Geyser-like Plumes: Discovery and Basic Characterization." *Science*, vol. 250, pp. 410–415. A technical account of eruptive plumes discovered during the *Voyager* 2 mission to Triton.

Thorpe, A. M. 1989. "Enigmatic Triton and Nereid." *Sky and Telescope*, vol. 77, pp. 484–485. Provides a clear explanation of Triton's decaying orbit.

Diamonds Are Forever

The Discovery of Stardust in Meteorites

Once during my childhood, while searching for seashells on a sandy Florida beach, I found a diamond. It was only a small diamond, a tiny chip really, set in an inexpensive ring, but I was thrilled with the notion that I had found something so rare as this sparkling gem. The ring itself was rather badly corroded and wrapped in seaweed, as if it had spent some time in seawater before being washed ashore. The glittering faceted stone, however, was untarnished by its residence in the ocean.

Diamonds are not indestructible, but they are impervious to most of the everyday processes that slowly erode and destroy more conventional materials. Diamonds will scratch any other substance and thus are assigned the ultimate value of ten in Moh's scale, the relative measure of mineral hardness. Antique diamonds do not lose their facets or their sparkle. Perhaps because they are so unchangeable, these stones have become the symbol of love everlasting. To borrow a phrase from diamond brokers' television commercials, diamonds are forever.

The formation of diamonds normally requires tremendous pressures. They form hundreds of kilometers deep within the Earth's mantle, as carbon atoms are packed into a densely compressed crystal structure. The diamonds thus formed are rafted to the surface in *kimberlites*, explosive pipes of magma that somehow manage to punch through great thicknesses of overlying rock. Successful attempts to synthesize diamonds in the laboratory for industrial purposes employ pressure vessels that squeeze the carbon atoms in the same way that nature does in the bowels of the Earth. A few meteorites also contain diamonds formed by shock waves when these projectiles slammed into the Earth or other orbiting bodies in space. None of these ways of forming diamonds is really commonplace, so the glittering crystals that result are rare and valuable.

Yet another, even more miraculous source of diamonds was discovered just a few years ago. This source has produced countless trillions of diamonds, but not a single one will ever grace a pendant or seal a marriage vow. These diamonds are stardust, tiny motes of solid matter created in doomed stars and ejected into the relative emptiness of interstellar space.

CHONDRITES

The stardust diamonds were discovered in a class of meteorites called chondrites, named for the Greek word *chondros* meaning "grain" or "seed." The name alludes to the distinctive textural appearance of chondrites. These meteorites commonly contain many small rounded objects, called chondrules, which are about the size of shot pellets. Perhaps the best description of chondrules was given by H. C. Sorby, a nineteenth-century English gentleman with a passion for geology and a large personal bank account that allowed him the luxury of pursuing this hobby. Sorby is famous as the inventor of the petrographic microscope, a tool with which geologists examine paper-thin sections of rock to identify their constituent minerals and to gather textural clues about their origin. When Sorby first peered at a chondrite through his microscope, he exclaimed that it contained "droplets of a fiery rain." The chondrules that got Sorby so excited contain crystals embedded in glass, a clear indication that they were once molten droplets. Recognizing that the formation of chondrule melts would require very high temperatures, Sorby suggested that they were

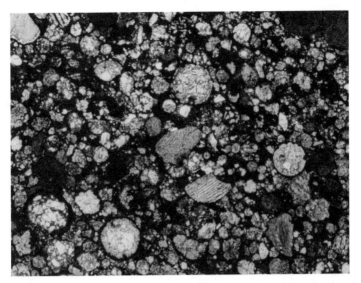

This magnified view of a chondrite that fell in Tieschitz, Czechoslovakia, is similar to what H. C. Sorby first saw through his petrographic microscope. The abundant rounded chondrules, embedded in dark matrix, are characteristic of this type of meteorite. Chondrites like this contain minute quantities of stardust.

pieces of the Sun ejected in solar flares. This quaint idea was probably prompted by the prevailing view at the time that the Sun was a rocky body wrapped in incandescent gases. Sorby's hypothesis is no longer tenable in light of modern ideas about the Sun's gaseous composition. More than a century later, however, we still have not unraveled the secret of how chondrules formed.

The meteorites that contain chondrules are arguably the most important and exciting matter available for scientific scrutiny. Chondrites are the only rocks that can be traced all the way back to the birth of the solar system. The age of chondrites, when determined from studying the gradual decay of their constituent radioactive atoms, is an astounding 4.5 billion years, older than any other known material. The Earth and other planets must have formed at about the same time as the chondrites, but the ages of terrestrial and lunar rocks are younger because their radioactive clocks have been reset repeatedly by later heating events. In fact, the 4.5-billion-year age of chondrites is normally used to define the solar system's birthday. Unlike planetary rocks, chondrites have mostly escaped the ravages of geologic processing. They are intact samples of the kinds of matter from which the planets must have been constructed.

Chondrites were made, at least indirectly, from the tiny bits of matter roaming the vast spaces between stars, the so-called interstellar medium. Some 4.5 billion years ago, interstellar dust and gas were gathered into the *solar nebula*, a cocoon that enveloped the newly ignited Sun during and shortly after its formation. This nebula was the factory where chondrites were assembled. Chondrites are basically mixtures of two ingredients—the spectacular droplets that we recognize as chondrules and a less distinctive, rather fine-grained material known as matrix. The chondrule droplets presumably formed by melting of interstellar dust in the nebula. Additional dust was vaporized and then recondensed as new minerals to form the matrix. The point is that the original interstellar dust and gas were transformed into second-generation minerals in the solar nebula; this once-foreign matter became the stuff of our solar system. Like immigrants assimilated into the culture of a new land, the particles of interstellar matter were homogenized and lost their distinctive identities when they moved into the solar neighborhood.

Once formed, these new minerals began to clump together, first as small orbiting clods and boulders, which in turn aggregated into larger "protoplanets" perhaps several hundred kilometers in diameter, finally culminating in massive planets. The present-day asteroid belt is a repository for leftover protoplanets that were not utilized in planet formation. Their preservation is a cosmic accident, a result of the fact that they happen to be located near Jupiter. The gravity of this massive planet would have wrenched apart any planet-size body that tried to form within the asteroid belt, so these few protoplanets escaped the fate of their kin, which were gathered into planets in other parts of the solar system. Most asteroids themselves are too small to have experienced much geologic processing, so they still retain their primitive characteristics. Collisions between asteroids occasionally knock off small pieces, a few of which eventually fall to Earth as chondritic meteorites.

CHEMICAL FINGERPRINTS IN CHONDRITES

The first hint that chondrites might actually contain some stardust—that is, unprocessed interstellar matter that somehow survived the solar nebula crucible—came from the analysis of gases in them. Helium, neon, argon, xenon, and krypton, the so-called noble gases, do not bond with other, presumably less exalted elements. Chondrites contain only tiny amounts of these gases, derived from many sources. Some noble gas atoms were accidentally trapped in minerals as they grew in the solar nebula. Atoms that are expelled from the Sun or other stars, called cosmic rays, are another source of noble gases in chondrites. Still more noble gases were created when certain radioactive atoms in the chondrites decayed. This is a long and tangled list of possibilities for noble gas sources, and the study of these gases in meteorites has evolved into a complicated specialty.

Of the various noble gases in chondrites, xenon has proved to be the most interesting. William Ramsay, who discovered the element xenon in 1898, named it for the Greek word meaning "stranger." That was a prophetic choice. In a sense, all of the elements in the solar system are foreign-born, but xenon is one of the few that has retained its foreign accent, at least in chondrites.

Because xenon does not participate in reactions with other elements, you might think that its chemistry would be simple. Studying xenon in meteorites, however, is not for the faint-hearted. The complexity of this element lies in its isotopes. The nine isotopes of xenon contain different numbers of neutrons and, consequently, different masses. (Atomic mass is the sum of neutrons plus protons.) Each batch of xenon formed in a particular environment has a distinctive mix of isotopes, providing a diagnostic isotopic fingerprint of the process that formed it. The xenon produced during each process can be separated through stepwise heating of a chondrite to progressively higher temperatures in the laboratory. With luck, the gases formed by each process emerge one at a time, as the host mineral for each one decomposes at a certain temperature.

In 1964 John Reynolds and Grenville Turner of the University of California at Berkeley carried out just this kind of experiment. While heating a chondrite in progressive temperature steps and analyzing the xenon isotopes given off during each step, they found not only the expected parcels of xenon, but two more they had not bargained for. One of these surprise parcels, which they labeled xenon-H, was enriched in the heaviest isotopes of xenon; the other, enriched in the lightest xenon isotopes, was named xenon-L. Scientific attention was mostly directed to xenon-H, because its peculiar mix of isotopes resembled that produced by the fission, or breaking apart, of uranium. Since the meteorite did not contain enough uranium to account for xenon-H, a mad search began for some other unstable element, probably now all decayed, that might have been its parent. This now-extinct parent element was said to be superheavy, as it had to be more massive than uranium (which is already in the sumo-wrestler category as elements go). A few superheavy elements such as fermium and nobelium have been made artificially in the laboratory, but they are so unstable that they exist for only a fleeting instant before decaying. The search for a superheavy element in meteorites was an attempt to see if such elements had ever formed naturally.

Further analyses of xenon confirmed that the same strange xenon-H and xenon-L parcels which Reynolds and Turner had discovered were present in other chondrites. One unexpected observation, however, was that xenon-H and xenon-L always occurred together in the same relative proportions in different meteorites. If you added the same unknown spice to various dishes, which always turned out both bitter and salty, you might assume that both effects were caused by this one spice. Using the same logic, researchers came to realize that xenon-H and xenon-L were actually a single component, which was renamed xenon-HL. The peculiar composition of xenon-HL, enriched in both the heaviest and the lightest isotopes, clearly could not be explained by fission. This result was somewhat embarrassing for champions of the superheavy element idea, who then had to abandon their search.

So what was the origin of this unusual xenon-HL in chondrites? A number of ideas were proffered, but of course no one really knew. None of the solar system processes we knew about seemed capable of producing this exotic xenon spice, so a controversial idea that it might be interstellar matter was bandied about. But one feature of xenon-HL seemed indisputable: It must be contained within a solid mineral in chondrites. Remember that xenon is a gas, so xenon-HL would have already mixed with other xenon in the solar nebula if it had entered the solar system in gaseous form. The very fact that this isotopically distinctive batch of gas survived in recognizable form means that it was carried into the nebula within a solid host, sort of a mineralogical equivalent of Typhoid Mary. Could chondrites really contain unaltered interstellar solids that were tagged with xenon-HL? Perhaps the origin of this foreign xenon would become clearer if the carrier grains could be isolated and identified.

SEPARATING STARDUST

The separation of stardust from chondrites, of course, is no mean feat. Imagine shaking a small amount of salt into a bubbling pot of vegetable soup, stirring vigorously, and then trying to retrieve the salt grains! If you have ever oversalted your food and tried to rectify this mistake, you know how difficult this is. You could try concentrating the salt by first filtering out the vegetables, on the assumption that all the salt dissolved in the liquid. This assumption is testable by tasting the broth to see if it is saltier than the original soup. You could then force the salt to precipitate by evaporating the remaining liquid, but the resulting salt grains would probably be mixed with some other previously dissolved ingredients. As a final step, you might have to pick out individual salt crystals by hand using a magnifying glass and tweezers. In any case, you could ostensibly trace the salt's trail by tasting each step in the separation. In an analogous way, xenon-HL can be used as a tracer for the separation of interstellar dust in chondrites, although the tasting procedure for xenon-HL is much more cumbersome and exacting.

The isolation of interstellar dust in chondrites is an interesting story in itself, a monument to scientific perseverance. Chemist Edward Anders and a succession of his colleagues at the University of Chicago worked off and on for nearly twenty years to separate and identify this material. But, as we will see shortly, the prize was worth waiting for.

Initially, Anders himself admits, they did the right experiment for the wrong reason. On a mistaken hunch that the mineral carrying xenon-HL was a sulfide, they dissolved a chondrite sample in acids that they knew would not affect sulfides. What remained was a dark black residue, consisting of sulfide plus several other minerals and representing only about one-half of 1 percent of the meteorite. This residue contained most of the xenon-HL, but to their consternation they soon discovered that the gas was not present in the sulfide. Next, they laboriously separated each of the other minerals in the black residue and tested them for xenon-HL, finally settling on carbon as the noble gas carrier. At this point, only about one-thousandth of the original meteorite mass was left, but even this was not a pure sample of the carrier. The interstellar carbon dust remained hidden among hundredfold larger amounts of garden-variety, solar system carbon that contains no xenon. Further laborious steps finally produced primal interstellar grains. In the process of obtaining this prize, Anders and his colleagues had dissolved away virtually all of the chondrite, as the interstellar dust comprised only four parts in ten thousand of the original stone's weight. If the search for interstellar dust in meteorites was like looking for a needle in a haystack, the approach taken was akin to burning the haystack.

During the final removal of unwanted solar system carbon, the residue dramatically changed its color from black to white. What form of carbon is white? In the ensuing excitement, Anders' research team drew on every sophisticated analytical technique at its disposal to characterize the remaining traces of mysterious white powder. What they had isolated was a minute amount of diamond dust. The individual xenon-laced diamonds are very tiny, averaging only twenty-six angstroms (about one ten-millionth of an inch) in diameter. Diamond crystals of this size consist of only a few thousand tightly packed carbon atoms. Anders described the size of the tiny grains this way: "If bacteria had engagement rings, these diamonds would be about the right size."

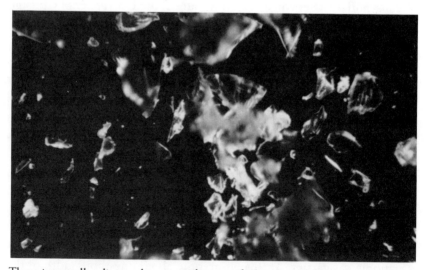

These interstellar diamonds are greatly magnified under an electron microscope. The individual crystals are extremely small and many have a yellow-brown color, owing to the presence of trapped nitrogen and other gases. These tiny particles formed around another star. (*E. Anders, University of Chicago.*)

The isolation and identification by Anders and his coworkers of tiny grains forged in another star was a riveting discovery. Who would have thought that stardust would consist of glittering diamonds? Even composer Hoagy Carmichael, who titled his best-known romantic ballad "Stardust," could not have dreamed up anything quite so fitting.

But stardust is not just diamonds. Many different minerals must have entered the solar system originally as interstellar dust, but only a small handful, other than diamond, have now been recognized in chondrites. One of these is silicon carbide, known commercially on Earth as carborundum. Like diamond, the tiny grains of silicon carbide carry their own distinctive noble gas fingerprints, equally bizarre but different from xenon-HL.

It is interesting that diamonds and silicon carbide find the same industrial uses on Earth. Both are extremely hard substances and make fine abrasives. It is the toughness of these interstellar grits that let them survive the harsh chemical treatments that dissolved the rest of the host chondrite. Whatever solid remains in a test tube full of hot acid represents "survival of the fittest." Diamond and silicon carbide are not the only kinds of interstellar grains in meteorites, but they are perhaps the hardest to destroy. Using a gentler, more time-consuming procedure, Anders has subsequently managed to isolate interstellar grains of graphite, a soft, friable form of carbon, familiar to most people as pencil lead.

WHERE STARDUST FORMS

These tiny grains clearly formed somewhere outside the solar system, as evidenced by the observation that they are tagged with noble gases whose bizarre isotopic compositions cannot be explained by any conceivable solar system process. At one time, these grains must have traveled between the stars, finally arriving in the vicinity of our own star at approximately the time of its birth. It is highly unlikely, however, that these grains formed in the near vacuum of interstellar space.

Stars provide a more suitable cauldron in which to forge diamond crystals. These tiny sparklers may have formed in gas expelled from a massive star. As the gas fled the star and cooled, carbon vapor could condense into solid form. But would this solid carbon really be diamond? We already have seen that this mineral forms deep in the Earth's interior or by shock during meteor impacts because extremely high pressures pack the carbon atoms tightly together. The pressures in expanding stellar gas envelopes are thought to be many millions of times lower than those required to make diamonds. In the last few years, however, Russian and Japanese scientists have managed to synthesize industrial diamonds at low pressures. This commercial process breaks down methane (a simple compound of carbon and hydrogen atoms) in the presence of hydrogen gas to form diamonds. The conditions necessary for the formation of diamonds by this process are thought to be similar to those in gas shells expelled by stars. What humans can do, apparently nature has done before.

To see how a star might shed its substance and thus form diamond stardust, we must consider how it evolves over time. When hydrogen atoms are heated to high temperatures and held together by the enormous pressures in the core of a star, the atoms fuse together to make helium atoms. Within our own Sun, some 400 million tons of hydrogen are converted to helium in this way every second. This process, called hydrogen burning, supplies most of the star's energy. Photons produced as by-products of the reaction work their way outward to the surface over a million years or so and escape into space, producing the star's visible light. When the hydrogen fuel begins to run low (which thankfully will not happen in our Sun for some time), a second fusion reaction occurs in which helium atoms, the ashes from the previous burning cycle, are themselves combined to make carbon and oxygen atoms. While helium burning is taking place in the core of the star, a hydrogen-burning shell works its way outward toward the surface; a star at this evolutionary stage expands to giant size and cools to a reddish color. The distended red giant becomes rather carbon-rich when enough helium has been burned, and it is just this kind of star that might produce diamond dust. A red giant frequently sloughs off gas shells, because its outer surface is so far removed from the center of the star that its gravitational hold becomes feeble. An expanding gas shell could then cool and condense into a cloud of diamonds.

The conditions necessary to make silicon carbide and graphite suggest that these interstellar grains, too, formed in the gas shells of carbon-rich, red giant stars. However, the noble gas isotopes in them vary from grain to grain, suggesting that perhaps a dozen different stars may have contributed to the assortment of silicon carbide and graphite grains found in chondrites.

STELLAR EXPLOSIONS

The unusual xenon contained within interstellar diamonds tells a fascinating story. Within the interior of a bloated red giant, continued fusion of atoms heavier than helium will produce not just carbon but a succession of even heavier elements, as the ashes from one burning stage become the fuel for the next. These stellar fusion reactions ultimately can build elements as heavy as iron, but that's about it. They cannot make gold, as the alchemists dreamed, nor can they make the peculiar mixture of xenon isotopes characteristic of xenon-HL.

The nuclear processes required to make xenon-HL occur only in supernovae, the brilliant explosions of stars. Near the end of a massive star's evolution, its nuclear fuel for fusion is nearly spent. At this point its interior is rapidly transformed into iron. When the iron core reaches a critical size, it collapses on itself and then violently rebounds, tearing the star apart. During this brief moment, much unique nuclear chemistry takes place, resulting in reactions that are not possible in any other setting.

The xenon-HL in interstellar diamonds is enriched in heavy and light isotopes of xenon that can only be formed, as far as we know, during the split second of a supernova. An old red giant could undergo such an explosion when it finally exhausts its supply of nuclear fuel. An explanation for diamond stardust, then, is that a carbon-rich red giant expelled a gas shell, which cooled and condensed as diamonds. Then, perhaps hundreds to a few thousand years later, this same ruddy star finally exhausted its nuclear fuel and exploded. Xenon-HL produced during this event was scattered outward at high velocity by the explosion. The expelled xenon atoms eventually overtook the more slowly expanding cloud of diamond particles, and some xenon-HL was implanted into the tiny crystals. This dust, now tagged with noble gas atoms produced during the supernova, drifted our way and ultimately became part of the assortment of interstellar dust and gas in the solar nebula.

Two superimposed telescopic photographs show a negative (dark) image of stars in the Andromeda galaxy and a positive (bright) image of this region after the 1987 supernova. The black dot at the center of the expanding cloud is the star that exploded. (*D. Malin, Anglo-Australian Telescope.*)

Just such a stellar explosion startled an astronomer at an observatory in Chile on a February evening in 1987. Comparing a new photograph of the Andromeda galaxy with one taken the night before, Ian Shelton noticed that something was wrong. There was a new object, a very bright object, where before there had been only an ordinary star. Shelton rushed to tell the crew at the telescope, and everyone went outside to gawk.

Death had come quickly to this star, and astronomers around the world were morbidly gleeful at its demise. This was, after all, the first supernova observed since the invention of the telescope. Of course, the stellar explosion had not just happened. It took 170,000 years for its light to reach us from the Andromeda galaxy, but for Earthbound astronomers it was as if they were watching in real time. In its death throes the star emitted light at a rate 100 million times greater than the Sun. High-resolution images indicated that the star had already shed its outer layers by the time of the supernova, and an expanding gas cloud blown away by the explosion was racing toward the shells expelled earlier. Although it is tempting to speculate that these shells contain specks of diamond, and perhaps by now they may be impregnated with supernova xenon-HL, this is probably not the case. The star that exploded was not a carbon-rich red giant, but a smaller blue giant with a somewhat different composition. However, as one astronomer put it, a blue supergiant is "the same bomb in a different suitcase." Red giant supernovae are represented by numerous ghostly remnants, such as the Crab Nebula, that spewed stardust into the interstellar medium thousands or millions of years ago.

The interstellar grains in chondrites, then, are the grime of a dozen other stars and the smoke of at least one supernova. Myriads of such interstellar particles must have been recycled ultimately to form a second-generation star, our Sun, and its attendant solid matter, the planets. I suppose it is not surprising that dust from many previous stars was used to construct our solar system, and in hindsight it may seem unusual that the interstellar diamonds in chondrites appear to be derived from only one supernova. Perhaps there are many more red giants belching gas clouds that form silicon carbide and graphite than there are supernovae. If that is true, then it is puzzling that diamond dust is much more abundant than silicon carbide and graphite dust in chondrites. Perhaps the supernova that produced the diamonds was closer to us than the many stars that provided silicon carbide and graphite dust to the nebula. Some researchers have postulated that a nearby supernova may have triggered the formation of the solar system, as its advancing shock wave compressed interstellar gas and dust into a tight bundle that ultimately became a star. Observations of supernova remnants show that new stars are embedded in their expanding clouds. Perhaps interstellar diamonds in chondrites are fossils of such a cosmic event.

WHAT STARDUST REVEALS

When the stardust formed is anybody's guess. What is clear, however, is that a tiny fraction of the matter in chondrites is older than the solar system itself, probably older than anything else we will ever get our hands on. Perhaps diamonds really are forever.

The importance of stardust in meteorites, however, transcends its mere antiquity. Stellar astronomers and astrophysicists have crafted an entire scientific discipline using telescopic observations and an understanding of the laws of physics. From rather meager information they have proposed wondrous processes that supposedly take place within the interiors of red giants and other stars, and during supernova explosions. But are these ideas right? Perhaps stardust can tell us.

These tiny motes of diamond and other minerals are etched with an indelible chemical memory of nuclear reactions in stellar interiors that we can speculate about but never actually observe. There are few places in science where one has the potential of drawing such grand conclusions from the analysis of such a paltry amount of material.

At the moment of his discovery of champagne, Dom Perignon is reported to have exclaimed to his fellow winemakers, "Come quickly, I am tasting stars!" In a sense, he was right. Bubbly champagne, as well as the people who enjoy it and the planet on which they live, are made of atoms that were created in stars. However, champagne, people, and planets are not stardust. Their constituent atoms have been repackaged repeatedly into new compounds or new minerals by biologic and geologic processes in our solar system. Only within the ancient chondrites can we find a few remnants of stellar material in its original form. In isolating and studying these little bits of matter, scientists have begun to savor the flavors of the cosmos and, in the process, to learn how the stars shine.

Some Suggestions for Further Reading

Anders, E. 1988. "Circumstellar Material in Meteorites: Noble Gases, Carbon and Nitrogen." In *Meteorites and the Early Solar System*, eds. J. F. Kerridge and M. S. Matthews. Tucson: University of Arizona Press, pp. 927–955. An authoritative but rather technical review of the discovery and analysis of interstellar grains in meteorites.

Beatty, J. K. 1987. "Stardust on Earth." *Sky & Telescope*, vol. 73, p. 610. This is an especially well written account of the stardust that resides in meteorites.

Lewis, R. S., and Anders, E. 1983. "Interstellar Matter in Meteorites." *Scientific American*, vol. 249, no. 2, pp. 66–77. An interesting account of the kinds of interstellar matter in meteorites; now somewhat dated in light of more recent discoveries.

Wood, J. A. 1979. *The Solar System*. Englewood Cliffs, NJ: Prentice-Hall. Chapter 6 of this excellent book gives a thoughtful and nicely illustrated description of stellar nucleosynthesis.

Woosley, S., and Weaver, T. 1989. "The Great Supernova of 1987." *Scientific American*, vol. 261, no. 2, pp. 32–40. Two leading astrophysicists describe supernova 1987A, the first such stellar explosion in more than 300 years.

Bull's-Eye

Multiring Basins and Cataclysmic Impacts

The greetings from Uncle Sam directed me to my first military duty station, an out-of-the-way air force pilot-training base in the west Texas desert. I had eagerly volunteered for this duty, despite some qualms about the dangers inherent in piloting high-performance aircraft. I had always marveled at how pilots responded coolly and flawlessly to in-flight emergencies and was concerned about not having the total recall ability they seemed to possess. Within the first few hours of flight instruction, their secret became apparent: The military has an obsession with checklists, which detail the proper responses to every emergency flight situation. During the subsequent months of training, we fledgling pilots practiced the slavish use of checklists while strapped into flight simulators, responding to whatever mechanical nightmares the instructor could devise. During my first simulator exercise, I heroically extinguished an engine fire by calmly and carefully accomplishing every item on the appropriate checklist. Anticipating a nod of approval, I glanced at the instructor. With a devilish smile he called to my attention the annoying buzzer in my helmet, an indication that the simulated aircraft had already crashed and burned several minutes earlier. Maybe the checklist should have contained a warning that it is necessary to continue flying the airplane while reading the checklist.

Like me, you may have discovered, at one time or another, that you focused so intently on solving some small problem that you lost sight of some larger purpose or context. It's a common response—we often don't see the forest for the trees.

Of course, that could not happen to trained and disciplined observers, such as planetary geologists. Or could it? How else could lunar scientists have missed *multiring basins*, the largest geologic structures on the surface of the Moon? These structures have been readily visible using small telescopes since the seventeenth century, but their very existence was not recognized until the middle of this century and was not widely appreciated until a decade later. This is even more surprising when you consider that the distinctive shapes of these features—gigantic bull's-eyes—call attention to them. The constituent parts of these bull's-eye patterns had been carefully mapped and described by generations of scientists, who were apparently focusing so much on the details that the big picture eluded them.

IMPACT BASINS ON THE MOON

Multiring basins are the largest impact features in the solar system. The collisions that produced these basins are described as catastrophic, but that term hardly does justice to effects so immensely destructive that we can hardly imagine them. Each collisional castastrophy leaves behind a pattern of concentric rings, formed by alternating mountain chains and valleys. When viewed from far enough away, these ridges and valleys resemble the black-and-white rings of a dart board. On the Moon, the centers of multiring basins are commonly filled with huge dark patches, the maria. These sound like features that would be hard to ignore, don't they?

The first person to describe lunar multiring basins was Ralph Baldwin, in an important book entitled *The Face of the Moon*, published in 1949. Baldwin had become interested in craters while serving as a bombadier during World War II, and after this experience he shifted his sights to craters not of his own making. His book marshaled arguments that craters on the Moon were caused by meteor impacts. Unlike many of his predecessors, he did not dwell on the interpretation of fine observational details but stressed large lunar surface features. Baldwin noted that the maria were roughly circular and were surrounded by multiple rings of mountain peaks. He also saw radial valleys, ridges, and grooves that are arrayed around the maria as other facets of the same process that produced the concentric rings. Others had observed these features in isolation, but Baldwin was the first to bring them together under the banner of a common origin. In doing so, he came to the surprising but inescapable conclusion that the mare lavas had filled gigantic impact craters. Moreover, the repetition of similar scars around every mare suggested that the Moon's nearside had been shaped by at least half a dozen massive impacts.

Multiring basins are much too large to be seen within the field of view of a typical telescope used by astronomers to map the details of the lunar surface, which may partially explain why Baldwin's predecessors failed to recognize their existence. In addition, the multiring basins on the lunar nearside are relatively old features, now pockmarked with numerous smaller craters. They are like bull's-eyes that have already been used many times for target practice, and by a poor shot at that. You may have seen dart boards so riddled with holes that they were difficult to score correctly. The distinctive patterns also have been covered in places by later lava flows, which spilled out where the circular rings were breached, or by overlapping ejecta blankets. Baldwin could see the bull's-eye pattern through all this clutter in his mind's eye, but most people could not. The view of a pristine, relatively unbattered bull's-eye was needed before the existence of these titanic structures could be widely appreciated.

Just such a bull's-eye was discovered in 1961 by Gerard Kuiper and his students at the University of Arizona. Kuiper had been engaged by the Air Force Aeronautical Chart and Information Center to prepare new maps for portions of the Moon. Part of this project involved the production of a rectified lunar atlas, showing the barely glimpsed edges of the Moon as they might appear if viewed from directly overhead. From different observatories Kuiper had obtained oblique but crisp photographs of the lunar limb regions, which he then focused through a large projector onto a three-foot-diameter white globe. By walking around to the sides of the globe, it was possible to view in almost ordinary perspective these regions that had only been seen before in extreme foreshortening.

One of the first features explored by this new technique appears, at least in the oblique view we get from Earth, as a modest dark patch nestled obscurely on one limb of the Moon. This feature, called Mare Orientale, was actually the center of an enormous multiring structure with elegant symmetry. William Hartmann, one of Kuiper's students, first suggested the term basin be applied to the magnificent Orientale structure and others similar to it.

A true overhead view of the Orientale multiring basin, as seen from a *Lunar Orbiter*, shows three prominent mountain rings. The spacing

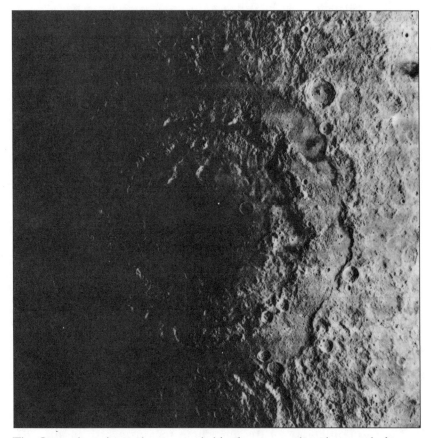

The Orientale multiring basin is probably the most striking large-scale feature on the lunar surface. Three concentric mountain rings form the most visible parts of a gigantic impact scar, the center of which was later filled by a vast lava lake called Mare Orientale. *(L. Kosofsky, National Space Science Data Center.)*

between the rings increases outward, a characteristic common to other multiring basins as well. The inner ring is a series of individual peaks rather than a continuous line. The second ring, which is the most prominent one, is a steep, continuous scarp. The intervening region between the second and third rings has a smooth appearance, and is thought to consist of materials that melted during the impact. The outer ring is a scarp that rises sharply 4 kilometers above the smooth terrain. The diameter of the outer ring is 900 kilometers, within which all six New England states would fit comfortably. Lying outside the last ring is a blanket of material ejected from the central cavity, now sculpted by radial grooves and chains of secondary craters.

The dark, flat center of the basin is Mare Orientale itself, a modest-size lava lake, at least by mare standards, that has sloshed over the crater floor. Like most other lunar basins, the interior of Orientale contains an abundance of dense rock, an extra concentration of mass relative to the crust around it. This feature, called a mascon in the vernacular of lunar scientists, was first detected by variations in the orbits of spacecraft zooming low over the surface. The distance of an orbiting vehicle above the lunar surface is controlled by its velocity and by the mass of the Moon. However, any increase in the density of the rocks located immediately beneath the spacecraft will modify its orbit slightly, because that will cause a tiny shift in the strength of the Moon's gravitational attraction at that point. If you were standing directly above a mascon, your body too would be pulled toward the center of the Moon by a slightly stronger gravitational force. You would not sense this tiny increase if you were walking about on the surface, but you would definitely weigh more as you hiked over a mascon. An orbiting spacecraft similarly is pulled more strongly toward the Moon as it passes over a mascon, causing a slight downward dip in its orbital trajectory and a corresponding increase in its velocity. Sudden accelerations have been detected in orbiters flying over all mare-filled lunar basins. The mare lavas are denser than those rocks surrounding the crater, so congealed lava lakes several kilometers thick may produce mascons. However, the lava pile alone may not account for the entire effect. Some scientists argue that mascons also require dense rocks from deep in the lunar interior to have been brought close to the surface under the lavas in Orientale and other multiring basins.

Photographs from lunar-orbiting spacecraft have revealed multiring basins in roughly equal numbers on both the front and back sides of the Moon but, curiously, lava lakes fill only those on the side facing the Earth. This phenomenon is apparently related to differences in crustal thickness. For reasons not well understood, the crust on the lunar farside is almost twice as thick as that on the nearside. This means that impact basins on the farside would not have excavated as deeply through the crust as those on the nearside. Batches of magma that formed in the deep lunar interior presumably rose high into the crust everywhere, their level of ascent dictated by their buoyancy. Although these magmas may have risen to about the same level on both sides of the Moon, that height was above ground level only at the centers of multiring basins in the thin, nearside crust. This situation can be likened to natural spring ponds on the Earth, in which pools of water appear in low areas that lie at or below the water table.

HOW MULTIRING BASINS FORMED

The origin of the repeated rings in basins has always been a controversial topic. The amount of energy expended in the formation of these basins is so great that it is difficult to imagine its effect on a rocky target. The kinetic energy of impact during the formation of the Orientale basin, if converted into electricity, would have been many thousands of times the current annual electrical output of all the Earth's civilizations, deposited at just one spot on the lunar surface within a time span measured in seconds or minutes.

Ralph Baldwin has been a longtime advocate of the idea that the target rocks at a gigantic impact site behaved like liquids. This may seem to fly in the face of everyday experience; everyone knows that rocks are hard and brittle. However, under the high temperatures and crushing pressures of a huge meteor impact, rock might deform in waves. Baldwin envisions the impact event to have looked something like a pebble falling into a quiet pool of water. The water does not form a permanent crater, of course, but it may illustrate in a simplified way what might happen to rock in a huge impact event. After the initial collision, a wave forms on the surface and expands outward. At the next instant, a peak rises in the center of the expanding crater.

After reaching a certain height, the peak collapses, producing yet another wave, which begins to move outward, and so on. This process repeats itself until friction damps out all the motions. Based on this analogy, the expanding waves of fluidized rock produced during the formation of a lunar basin might conceivably have behaved like tidal waves on the Earth's oceans. Monstrous ocean waves, called tsunami, may travel thousands of miles before angrily venting their energy on coastal shorelines. Oscillating tsunami of dense rock would be slowed more readily, finally freezing into concentric mountain ranges. In Baldwin's frozen tsunami approach to making multiring basins, the initial cavity was about the size of the inside ring, and the other rings propagated outward like ocean waves.

A competing theory suggests that the original crater rim was the outer ring. In this model, the inner rings formed as energy from the impact traveled downward into the Moon's crust and then was reflected back toward the surface. The reflection might have exposed underground rock layers within the crater, the upturned ends of which formed the interior rings.

In yet another model, the multiple rings formed much later than the crater itself. Multiring basins are proportionately much shallower than small craters, which probably means that they have been modified in some way after their formation. The outer rings could be huge blocks that slumped downward and inward along steep faults. The slumped blocks produced a series of terraces that stairstep down to the crater floor.

IMPACT BASINS ON OTHER BODIES

Had fate positioned the spectacular Orientale bull's-eye on the Moon's nearside, rather than hidden away on a limb, we would not have required the observational skills of Ralph Baldwin to recognize lunar basins. Viewed from Earth, Orientale would probably have looked like a huge, unblinking and omnipresent eye, the cornea of some stern, heavenly deity surveying his or her minions below. This impact scar might have spawned religious cults, dedicated to worshiping the god who watched over Earth at each full Moon. Perhaps it is lucky that we were spared this particular twist of nature.

With Baldwin's guidance and an example of a relatively pristine multiring basin on the Moon, planetary geologists trained their own eyes to see ringed basins, and have subsequently encountered similar gigantic craters on bodies throughout the solar system. One of the most impressive examples is the Caloris basin on Mercury. A calorie is a unit of heat, and calories are what the Caloris basin has a lot of. This basin is fortuitously situated at the point on Mercury's surface that receives the most direct illumination during the inner planet's closest approach to the Sun, where temperatures reach 350 degrees centigrade.

Caloris is a colossal feature some 1,300 kilometers in diameter, large enough to enclose the state of Texas. In late 1974 the *Mariner 10* spacecraft passed within 800 kilometers of the planet's surface and took a quick look at an area covering less than half of Caloris, but that was enough to recognize what this structure was. The perimeter of the basin is a rugged ridge rising 2 kilometers above the surrounding plains. One might quibble about calling this a multiring basin, as it has only this one visible ring. However, it seems likely that some rings have been covered by lava flows that smoothed the crater interior, or by deposits of material ejected from neighboring craters. We know that multiple rings can form during large impacts on Mercury, because a smaller basin with several rings has been observed on this planet. Outside the visible ring is a vast ejecta blanket, decorated with linear crater chains similar to those around lunar basins.

The impact that created the Caloris basin was so intense that it was felt all the way on the other side of Mercury. Centered on the exact opposite side of the planet from the basin is a bizarre terrain of hills and valleys more than 500 kilometers across. This peculiar region formed from a shock wave that traveled from the Caloris basin site through the planet and emerged on the other side.

Caloris and Orientale illustrate what happens when huge meteors collide with rocky targets. Of course, not all planets and satellites have rocky crusts. Callisto, the second largest moon of Jupiter, has a thick mantle of ice. Images of this densely cratered world sent back by *Voyager* spacecraft showed what are undoubtedly the most photogenic of known multiring basins, Valhalla and Asgard. The names of these basins are derived from the homes of gods in Norse mythology. Asgard is an intriguing feature, but Valhalla is clearly the star of the

The Caloris basin, measuring 1,300 kilometers across, is the largest geologic feature on Mercury. This composite photograph shows the western half of the basin. The interior is flooded with lavas, and a vast blanket of ejected material lies outside the main rim. (*Jet Propulsion Laboratory.*)

show. This giant basin is several thousand kilometers in diameter, dwarfing even the Caloris basin on Mercury. Valhalla has at least twenty-five separate concentric rings. The modifier "multiring" hardly seems adequate for this magnificent bull's-eye, and "ripple ring" has sometimes been used. The center of the basin is a highly reflective area, probably an exposure of cleaner ice from the moon's interior. In contrast to the basins on rocky planets, Valhalla is missing the textured ejecta blanket that surrounds Orientale and Caloris, and Valhalla's rings are evenly spaced. The stark contrast that Valhalla

This shaded relief map of part of Callisto's surface was compiled from Voyager images by the U.S. Geological Survey. Two spectacular multiring basins are visible on this densely cratered world. Large impacts into Callisto's ice crust produced basins with as many as twenty-five concentric rings. (*U.S. Geological Survey.*)

and its smaller companion Asgard offer to other multiring basins results from the different ways that ice and rock targets respond to impact. Ice flows more easily than rock, so perhaps the frozen tsunami model is a perfect metaphor for the formation of large craters on this world.

I have not alluded to multiring basins on the Earth, for the simple reason that there are none or, at least, none on which everyone can agree. Is there something special about the Moon, Mercury, and Callisto that would explain the occurrence of basins there?

Let's consider the rocky bodies first. The Moon and Mercury are so similar in appearance that a *Mariner 10* project scientist remarked that Mercury photographs could be substituted for lunar pictures and most people would not notice the difference. What makes the Moon and Mercury so similar to each other and so different from the Earth is that both have surfaces that are heavily cratered. An intensely cratered surface is a sign of old age, and the heavily pockmarked appearance of virtually all multiring basins means that they must be older still. Icy Callisto is likewise heavily cratered, testifying to the antiquity of its surface relative to the other nearby moons of Jupiter. The multiring basins on all these bodies are preserved, though damaged, because they happen to be on geologically dead worlds. These bodies lack the active geologic processes that continually remake their surfaces and destroy old features such as these huge craters. Multiring basins are in a way gigantic fossils, reminders of some ancient time thankfully past when the collisions of very large bodies were not unusual events. The Earth certainly did not escape this holocaust, but its former basins have been obliterated by the relentless surface processing of a geologically active planet.

BASIN-FORMING EVENTS

Although it is clear that multiring basins formed long ago, planetary geologists continue to argue about the significance of the timing of these massive impacts. Virtually everyone begins with the same assumption: The planets progressively grew to their current sizes by addition of matter through collisions of smaller bodies, called protoplanets. After that, their views diverge.

Multiring basins on the Moon formed between about 4.1 and 3.8 billion years ago. One group of scientists believes that a fearful lunar cataclysm occurred in this interval, when virtually all of the large impacts took place. They suggest that this time marks a point at which most of the orbiting matter had been collected into a few large protoplanets with incredible destructive power. For some reason, all of these bodies collided at nearly the same time, like billiard balls at the opening break. In their view, the formation of multiring basins marks the last stage of planetary formation.

Another group takes the position that collision rates have decreased steadily and progressively through time, as the solar system was gradually cleansed of orbiting debris. Protoplanets were swept up by the planets gradually, as billiard balls are sunk into pockets one at a time. In their view, multiring basin formation represents an era when impacts, both large and small, were commonplace but declining. This group argues that massive impacts also occurred prior to 4.1 billion years ago, but they did not form basins because the Moon was still partly molten. Only after appreciable cooling and thickening of the lunar crust could a permanent record of these impacts be made. Even if the crust had hardened before 4.1 billion years ago, previous basins were destroyed during the formation of later basins.

EVEN MORE DESTRUCTIVE IMPACTS

Multiring basins are the largest geologic features observed on a number of planets and satellites. These gigantic basins were produced by massive projectiles, ranging perhaps up to several hundred kilometers in diameter, impacting onto much larger bodies. As destructive as such collisions undoubtedly were, the net effect was still the complete destruction of the impactor and the permanent wedding of this material to the larger body. However, collisions between bodies of comparable size might have a very different consequence. In such a case, one or both bodies might be smashed into bits and pieces, so that mass would be lost or at least redistributed between new objects formed from these fragments.

Miranda, a tiny moon of Uranus, may provide visible evidence of just such a disruptive collision. Miranda takes its name from the heroine in Shakespeare's *The Tempest*, which in retrospect seems very appropriate. Although most experts predicted it would be a bland ball of ice, Miranda offered a few surprises when it was photographed by *Voyager* 2 in 1986. Its tempestuous surface has two distinctive regions, one heavily cratered like the Moon and one banded and folded into chevrons. These regions have sharp, angular boundaries, as if the whole moon is some improbable jumble of dissimilar fragments. A plausible explanation for Miranda's exotic appearance is that it was shattered into a number of large chunks by a giant impact. The

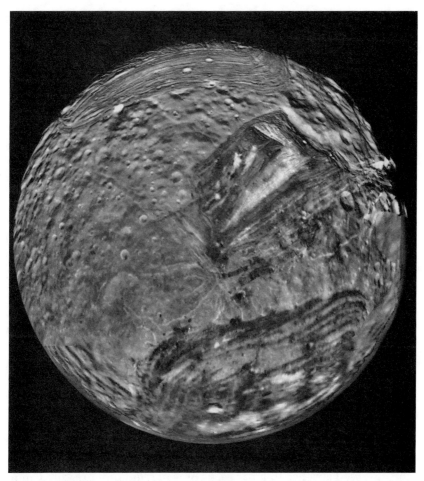

Miranda, a small satellite of Uranus, displays two contrasting types of geologic terrains, a densely cratered region and a layered terrain folded into chevrons. This bizarre moon may have been pieced together from the fragmented remains of two bodies, both of which were disrupted during a massive collision. The pieces were reassembled by mutual gravitational attraction into the object we now see. (*Jet Propulsion Laboratory.*)

fragments from this collision eventually reassembled themselves by mutual gravitational attraction into the collage we now see. Dislodged chunks from Miranda's interior and crust, possibly augmented by pieces of the body that pummeled it, were incorporated at random into the newly assembled moon.

The possible role of other, equally disruptive collisions in the evolution of planets has begun to be explored only recently. One currently popular and probably correct theory holds that the Moon originated when the Earth collided with a Mars-size impactor early in its history. The uniqueness of the Earth-Moon system has always been perplexing, and this idea goes a long way toward explaining its peculiarities. The Earth is the only one of the inner planets with a large satellite, the orbit of which is neither in the equatorial plane of the Earth nor in the plane in which the other planets lie. The Moon's mean density is much lower than that of the Earth but is about the same as the Earth's mantle. This similarity in density has prompted speculation, dating as far back as George Darwin (son of the famous evolutionist) in 1879, that the Moon split away from a rapidly rotating Earth. The idea founders on two observations, however. In order to spin off the Moon, the Earth would have had to rotate so fast that a day would have lasted less than three hours. It is thankfully not spinning that fast now, and we know of no plausible explanation of how it could have slowed to its current rotational rate. Moreover, the Moon's chemistry is similar to that of the Earth's mantle, but it is not a precise match. Theorizing a giant impact eliminates the necessity of having a whirling dervish Earth as well as offering a solution to the Moon's peculiar chemistry. In an impact model, the bulk of the Moon would have formed from a combination of material from the impactor and the Earth's mantle. Most of the Earthly component would have been in the form of melted or vaporized matter. The difficulty in recondensing this vapor in Earth orbit might account for the observed absence in lunar rocks of water and other readily vaporized compounds and elements.

Once we admit that such massive collisions could have occurred, it is tempting to let the imagination run wild and speculate about how similar impacts might have affected all the planets. For example, Mercury is known to have a high density in comparison with other planets. A titanic impact could have stripped away a portion of Mercury's rocky mantle, leaving behind a metallic core out of proportion with the original amount of rock. A massive, glancing blow to Venus might have slowed its spin and reversed its direction of rotation. The highly unlikely orientation of the spin axis of Uranus, which is lying on its side with its pole pointed toward the Sun, might likewise be explained by a massive impact.

These conjectures are all intriguing, but it is also disconcerting to consider that no planet in the early solar system was immune to the

This thought-provoking cartoon is a reminder that the Earth has not been immune to large impacts. *By permission of Johnny Hart and Creators Syndicate, Inc.*

ravages of monstrous impacts. Have we opened a Pandora's box of cataclysmic collisions that can be used indiscriminately to explain away every unusual planetary characteristic? Perhaps. But then again, we may just have begun to discern the big picture, a view even more expansive than that afforded by the recognition of multiring basins. After all, it is easy to miss the forest for the trees.

Suggestions for Further Reading

Baldwin, R. B. 1949. *The Face of the Moon.* Chicago: University of Chicago Press. This pioneering and very entertaining book contains the first observations of lunar multiring basins, as well as arguments for their origin by impacts.

Chapman, C. R., and Morrison, D. 1989. *Cosmic Catastrophies.* New

York: Plenum Press. A general-audience review of evidence for the emerging scientific debate on catastrophism. Of particular interest in the context of massive impacts are chapters on lunar craters and the Moon's origin, the extinction of dinosaurs, and nuclear winter.

Hartmann, W. K. 1977. "Cratering in the Solar System." *Scientific American*, vol. 235, no. 1, pp. 84–99. An easily understandable account of the mechanics of crater formation and the use of crater counts as a chronological tool.

Schultz, P. H., and Merrill, R. B., eds. 1981. *Multi-ring Basins. Geochimica et Cosmochimica Acta*, Supplement 15. New York: Pergamon Press. Technical papers dealing with research on multiring basins on the Moon, Mercury, Mars, and other bodies. The contributions by Grieve and coworkers, Hartmann, Croft, McKinnon, and Baldwin are especially interesting. Some papers are difficult reading.

Taylor, S. R. 1987. "The Origin of the Moon." *American Scientist*, vol. 75, pp. 468–477. An excellent summary of current scientific thinking on the Moon's origin by catastrophic impact on the Earth.

A Piece of the Red Planet

———

Meteorites from Mars?

One hot, humid Houston evening in the spring of 1977, as I was sharing a cool drink with my friend Edward Stolper in a cowboy bar near the Johnson Space Center, an astounding thing happened. While sitting in the dark surrounded by country music and long-necked beer bottles, we stumbled across several small pieces of the planet Mars. But I am getting ahead of myself. I had better backtrack and tell you the whole story.

At the time, Stolper and I were graduate students at Harvard. We had just collaborated on a study of several unusual igneous meteorites, formed from hot, molten magma. So, on that evening, Stolper and I were planning how we would cast this research into a technical paper. An outline for our paper was progressing quite nicely until we started talking about where these meteorites might have originated. At this point, Stolper mumbled something about how interesting it would be if they had formed on a large planet, say Mars. Well, I couldn't let that pass without an argument. I stubbornly marshaled all the reasons why this was an outlandish idea, and Stolper, who likes to argue as much as I do, countered effectively with his own points. Several beers later, we had both convinced ourselves that this highly unlikely hypothesis might just be possible. We were still too wary to publish this brash idea in our paper, but we did introduce it in a popularized *Scientific American* article submitted the following year.

SNC METEORITES

The notion that there could be even one Mars rock here on Earth is hard to swallow, but altogether we now have in our possession nine putative pieces of the red planet. *Shergottites*, the meteorites that Stolper and I studied, are basaltic meteorites, solidified lavas much like those that spew from Hawaiian volcanoes or comprise the floors of the world's oceans. The namesake for shergottites fell in 1865 at Shergotty, a small village in India, and a second meteorite of the same type plummeted into Zagami, Nigeria, almost 100 years later. Since the late 1970s, three more shergottites have been discovered on the Antarctic ice cap. The *nakhlites* take their name from a shower of about forty stones that pelted the village of El Nakhla el Baharia, Egypt, in 1911. One of these meteoritic projectiles reportedly struck and killed

This meteorite, a nakhlite, was recovered in Lafayette, Indiana. Its distinctive crust shows flight markings of surface material melted during atmosphere transit. The meteorite is about the size of a baseball. (*Smithsonian Museum of Natural History.*)

a dog. Individual specimens found in Lafayette, Indiana, and in Governador Valadares, Brazil, complete the nakhlite group. The only known *chassignite* startled the residents of Chassigny, France, in 1815. The nakhlites and chassignite are closely related to the shergottites, but each of these meteorite groups contains different minerals. Drawing on the first letter of each group, these stones are collectively referred to as SNC (pronounced as the individual letters "S-N-C" or as "snik," depending on your preference) meteorites.

The characteristic that sets the SNCs apart from all other kinds of meteorites is their age. By measuring the decay products of various radioactive isotopes in an igneous rock, it is usually possible to determine how long ago it solidified, which is the rock's *crystallization age*. Several independent clocks have been used to determine the crystallization ages of SNC meteorites, each chronometer based on the decay of a different isotope. Regardless of which isotopic clock is employed, the crystallization age of the nakhlites and the lone chassignite are the same, about 1.3 billion years ago. Exactly when the shergottites solidified is less certain, and estimates vary from 1.3 billion years ago to as recently as 180 million years ago. This uncertainty is due in part to the fact that shock caused by meteor impact may have reset some of the isotopic clocks. In any case, all the SNC meteorites clearly formed within the last 1.3 billion years, which makes them very young, at least by meteorite standards. Other kinds of meteorites have ages of around 4.5 billion years.

TENTATIVE ARGUMENTS FOR MARTIAN ORIGIN

The crux of our original argument that SNC meteorites were from Mars was their recent crystallization ages. We know of no way that magmas can be produced within small bodies billions of years after their creation. Even if small bodies started out hot 4.5 billion years ago, they would have cooled rapidly so that by 1.3 billion years ago their tepid interiors could not house molten rock. Only large planets appear to be capable of maintaining for aeons the high internal temperatures necessary for melting.

We can demonstrate by constructing a simple diagram that compares the size of each planet to the duration of its volcanic activity.

Along the vertical axis of this diagram, I have arranged the terrestrial planets and a few smaller objects by size, with the largest bodies, Earth and Venus, at the top and Vesta, a small asteroid only a few hundred kilometers in diameter, at the bottom. Vesta has been suggested to be the parent body for the *eucrites* (hence EPB, or eucrite parent body), another class of meteorites. The name eucrite comes from the Greek *eukritos*, meaning "easily distinguished"; the reason for this name is a puzzle, at least to me, because few geologists could readily distinguish chunks of eucrite from terrestrial rocks. In any case the eucrites, like SNC meteorites, are igneous rocks, but the eucrites crystallized much earlier, between 4.4 and 4.5 billion years ago. The eucrite parent asteroid was obviously a midget, much too small to retain heat for any appreciable length of time. In earliest solar system history, potent heat sources, such as compaction or the rapid decay of short-lived radioactive isotopes, may have caused melting of asteroids like Vesta soon after their formation. Heat from these sources quickly dissipated, however. Only larger planets with thick blankets of insulating rock were warmed continually by the slow decay of long-lived radioactive isotopes.

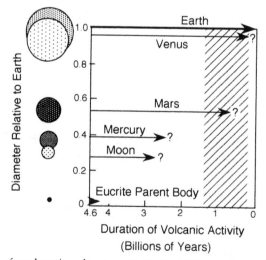

The duration of a planet's volcanic activity (in billions of years) is related to its size. The hatched vertical bar represents the crystallization ages of SNC meteorites, which may be martian samples. EPB indicates the eucrite parent body, possibly the asteroid Vesta.

Returning to our diagram, the horizontal axis is a time scale, ranging from 4.5 billion years ago, the approximate time at which all of the planets formed, to the present day. The duration of volcanic activity on each body is represented by an arrow stretching from its beginning to the time of the last volcanic eruption. No eucrites are younger than 4.4 billion years, so Vesta's volcanism was just a flash in the pan. The youngest rocks brought back from the Moon crystallized 2.8 billion years ago, so its arrow extends to that time. The ages of volcanic plains on planets that we have not yet visited and sampled, such as Mercury and Mars, have been estimated by counting the abundance of craters on their surfaces. We do not know whether Venus is now volcanically active because its surface is obscured continually by dense clouds. However, some volcanic structures imaged by the *Magellan* radar mapper look very fresh, and sudden changes in the composition of the planet's atmosphere suggest recent outpourings of volcanic gases. Obviously, the exact times that volcanism ceased on unsampled planets are somewhat uncertain, but estimates accurate to within half a billion years are probably sufficient for this purpose. The Earth, of course, sporadically experiences volcanic eruptions in the present day. Our diagram clearly shows that the larger a planet is, the longer is its volcanic lifetime. An exception to this relationship, which I have not illustrated in this graph, is Jupiter's moon Io. However, we have already seen that the heat powering Io's ongoing volcanic eruptions has a completely different origin unrelated to radioactivity.

The crystallization ages of SNC meteorites are indicated on the diagram by a hatched vertical bar, the width of which reflects the controversy in the ages of shergottites. This bar overlaps the times of volcanic activity on large planets, but not on smaller bodies. Of course, this observation does not specifically demand a martian origin for SNC meteorites, but it does suggest that they formed on some large planet. Stolper and I chose Mars over Venus because the approximate ages of its volcanic plains were known and because Venus' greater size would make it even more difficult to remove samples from its gravitational grasp.

A few scientists have posited that the SNC meteorites formed as impact melts, puddles of molten rock produced by the heat of meteor collisions. This suggestion, if correct, would allow the possibility of an asteroidal source for these young meteorites, because impacts could occur at any time. However, SNCs show no resemblance to impact melts found in terrestrial and lunar craters, so this alternative seems to have little to recommend it.

Almost all of the SNC meteorites are *cumulates*, rocks that formed by the accumulation of dense crystals that sunk through less dense magma and piled up at the bottom of a magma chamber. We base this conclusion on the observation that individual elongated crystals in these rocks are oriented in the same direction. If you were to drop a handful of such crystals into a water-filled aquarium, each crystal would sink to the bottom and come to rest on its side rather than standing precariously on end. In a similar way, the elongated crystals in shergottites were oriented as they settled onto the floor of a magma chamber.

Heavy crystals sink through less dense liquid because of gravity. Larger, more massive crystals settle faster than smaller ones, and this natural size-sorting mechanism produces layers of accumulated

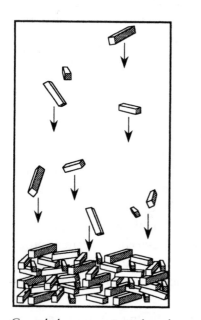

Crystals become oriented as they settle onto the floor of a magma chamber. The alignments of mineral grains in SNC meteorites were probably caused by this process. A slab of the Shergotty meteorite shows oriented pyroxene crystals.

crystals with nearly uniform sizes. Identical crystals will sink faster in a magma on a large planet such as the Earth than in that same magma on a small asteroid with a weaker gravity field. In 1981 Robert Grimm, then a student at the University of Tennessee, constructed a computer model to mimic the settling of crystals in shergottite magma. The gravitational force pulling the crystals downward can be counteracted, entirely or in part, if the magma is very slushy, as a result of its composition and temperature. Knowing the density and slushiness of the shergottite magma and the mass of the crystals, Grimm estimated the minimum gravitational force necessary to cause the mineral grains to sink. He concluded that a body of at least lunar size was necessary to produce the observed crystal settling in shergottites. The Moon has a diameter more than six times that of the largest asteroid, so lunar gravity exerts a force many times greater than any asteroid could. Grimm's study, coming as it did on the heels of the proposal that SNC meteorites were from Mars, fanned the flames by adding more support for a planetary origin.

A MORE CONVINCING ARGUMENT

Up to this point, all the evidence that had been gathered—the recent crystallization ages and the need for a sizable gravity field to cause crystal settling—merely suggested that SNC meteorites might come from a large planet. Although Mars had already been proposed as a plausible source, there was really nothing that specifically linked these meteorites to the red planet. Nothing, that is, until a straightforward experiment produced an unexpected result.

This particular experiment was designed to determine when the shergottites were ejected from their parent body and launched into interplanetary space. Some of these meteorites have been partly melted by shock, which resets the isotopic clocks in rocks. Therefore, analysis of the shock-melted glass in a shergottite should provide an estimate of the timing of the impact, which we will call the *shock age* to distinguish it from the crystallization age. In 1982 Donald Bogard and some colleagues at NASA's Johnson Space Center dug out a tiny, glassy chip from a shergottite melt pocket to measure its shock age. The particular clock Bogard used for this purpose was the decay of a radioactive isotope of potassium into argon. Impact melting should have flushed away the argon gas originally present in the sample, so that any measured argon formed by decay after the melt was quenched to glass. Knowing the rate at which radioactive potassium decays, it was then a simple matter to calculate the shock age. When Bogard did this calculation, he obtained an astounding result: The meteorite had been shocked 6 billion years ago! This result, of course, is absurd. The meteorite did not even exist then, because its crystallization age is no older than 1.3 billion years. In fact, the solar system itself did not form until 4.5 billion years ago. Something about this measurement seemed terribly amiss.

Faced with this experimental lemon, Bogard made a clever deduction. Clearly, there was far too much argon in the shergottite glass sample to have resulted simply from decay of radioactive potassium. He reasoned that extra argon, unrelated to potassium decay, had somehow entered the sample. But how and when was this accomplished? The key to this puzzle is the fact that argon is a gas. Like other gaseous elements, argon forms a portion of the atmospheres that surround some planets. Bogard suggested that a puff of air on the SNC parent planet was trapped within the shock-melted rock during the impact. He then estimated the isotopic composition of this trapped argon as well as the proportions of some other gases. The abundances of various gases and the proportions of different gas isotopes in the martian atmosphere had been measured six years earlier by Viking spacecraft. Lo and behold, the composition of this trapped gas precisely matched that of the thin martian atmosphere! The extraordinary match provided a compelling argument that SNC meteorites are from Mars. Who would have dreamed that a few whiffs of gases would provide a diagnostic fingerprint with which to link shergottite meteorites to the red planet?

METEORITES FROM THE MOON

The concept of martian meteorites initially met with general disbelief, if not downright hostility, from much of the scientific community. This response was predictable and justified, because, up to that point, no meteorites were known to have come from the Moon. If chunks of rock had never been torn from the lunar surface, how could they

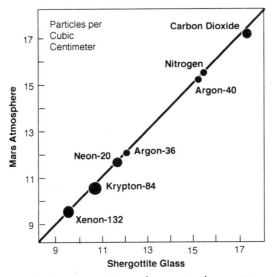

This comparison of the composition of gases in the martian atmosphere, as analyzed by *Viking* spacecraft, with trapped gas in glass pockets of a shergottite demonstrates that the two are remarkably similar. Shown are the amounts of carbon dioxide and nitrogen as well as various isotopes of argon, neon, krypton, and xenon. Both axes of the diagram are logarithmic, and an exact match would be indicated if all points fell along the diagonal line. The sizes of the dots indicate the amount of analytical uncertainty.

have been extracted from an even larger body? We champions of the martian meteorite hypothesis had no ready answer to this quandary, but one was shortly forthcoming.

Long ago, the notion that meteorites came from the Moon was not at all farfetched. In fact, it was one of the earliest explanations for meteorites, dating back more than three centuries. Just prior to the *Apollo* program in the 1960s, this idea was revived when Harold Urey, a highly respected meteorite researcher and Nobel laureate, suggested that eucrites were lunar meteorites. However, the direct comparison of eucrites with lunar samples returned by the *Apollo* astronauts proved this popular idea to be wrong, and the experience understandably left many planetary scientists gun-shy about accepting the possibility of obtaining meteorites from a planet even larger than the Moon.

As sometimes happens, a seemingly unconnected and faraway event changed all this. At the end of a bitterly cold, blustery day in early 1982, several members of an Antarctic expedition were traversing an ice field on snowmobiles. Visibility was poor because of blowing snow, but by chance a small black rock attracted the attention of one of the team. He stopped, examined the peculiar, golf ball–size object, and stuffed it into a collection bag. This particular Antarctic expedition had been mounted specifically to search for meteorites, and its members were experienced in distinguishing extraterrestrial stones from random bits of Antarctic bedrock. This small black rock, although apparently a meteorite, was unlike any they had ever seen or collected. A few weeks later it was shipped, along with the other meteorite booty of that field season, from McMurdo Station to the Johnson Space Center in Houston. Its destination was a facility formerly used to process Moon rocks during the *Apollo* program. Unbeknownst to the technicians who opened the package containing this frozen meteorite, the old Lunar Receiving Laboratory was back in business.

Part of NASA's standard procedure for processing newly recovered Antarctic meteorites is a preliminary examination, the results of which are published in a newsletter sent periodically to meteorite researchers all over the world. The job of examining and describing this particular meteorite fell to Brian Mason, a Smithsonian mineralogist with a wealth of experience in the study of meteorites and lunar rocks. The newsletter entry for this meteorite listed it as a breccia, a rock composed of broken pieces of other igneous rocks cemented together. Mason's guarded wording, to the effect that this meteorite breccia resembled fragmented lunar rocks, only hints at the excitement he must have felt on first examining this specimen. His subtlety was not lost on those who read the newsletter, however. His seemingly innocuous comment precipitated a flurry of sample requests, and by year's end no less than twenty-two research teams were hard at work on allocations of this meteorite. A special symposium to discuss this unusual meteorite was scheduled during the 1983 Lunar and Planetary Science Conference. Each and every symposium speaker in turn opined that this meteorite was a lunar sample. Such unanimity is unusual in the often-contentious world of planetary science. Since that time, a handful of additional lunar meteorites have been identified in the Antarctic meteorite collections of the United States and Japan, and one lunar meteorite has been found in Australia. Each one is different enough from the rest that the various specimens must have come from widely separated areas and were likely ejected from the

Moon in separate events. Apparently, the old idea that meteorites came from the Moon was not so farfetched after all.

Once the most cogent argument against the idea of a martian origin for SNC meteorites had been squelched, the advocates of this proposal breathed a collective sigh of relief. The discovery of lunar meteorites did not mean, however, that the rest of the scientific community was yet ready to jump on the bandwagon.

LAUNCHING MARTIAN METEORITES

How does a rock on the surface of Mars or the Moon become a meteor? The only plausible way to eject a rock from a planet's surface is to dislodge it during an impact. A meteor colliding with Mars or the Moon at a speed of several kilometers per second or more carries tremendous kinetic energy, most of which is used up in excavating a crater. Almost all of the rock fragments blasted outward from the crater eventually fall back to the planet's surface, but some tiny fraction of the target material might conceivably escape altogether. If the energy of motion of a lofted rock exceeds the gravitational pull that binds the planet together, the rock is said to have achieved "escape velocity." This term will be familiar if you have watched televised rocket launches. Martian escape velocity is five kilometers per second, twice that of the Moon but only about one-fourth that of the Earth.

A rocket accelerates to escape velocity gradually over a period of several minutes, but a rock hurled by impact from a planetary surface reaches escape velocity almost instantaneously. It seems likely then that such a rock would bear the scars of the intense shock that accompanied its launch. Surprisingly, the nakhlites and chassignite show only modest shock effects. The shergottites, however, were pounded fiercely and responded to the torturous pressures in several ways. Grains of feldspar were instantly transformed into glass, as the passing shock wave randomized the atoms within the crystals. Shock compression also caused portions of these rocks to liquefy, and the melt squirted through cracks and collected in small pockets, ultimately quenching to form more glass. The tiny pockets of impact-melted glass in these meteorites trapped the atmospheric gases and thereby provided a critical revelation about their home world.

Once this linkage between Mars and SNC meteorites had been

established, it was possible to speculate about specific locations that might be suitable sites from which to extract SNC-like rocks. There are actually few places on Mars where these meteorites could have formed, because most of the planet's surface is just too old. One part of the northern hemisphere of Mars is promising though. In this region are a number of gigantic volcanoes, including Olympus Mons, the largest such feature in the solar system. The volcanoes and the smooth lava plains around them have been estimated to be no older than 1.3 billion years, based on crater-counting. The largest impact craters in this relatively young martian terrain are about thirty kilometers across. Impacts that could dislodge sizable chunks of rock from the planet's gravity field would have to have been big, and probably created large craters like these.

Planetary scientists who study impacts were originally rather skep-

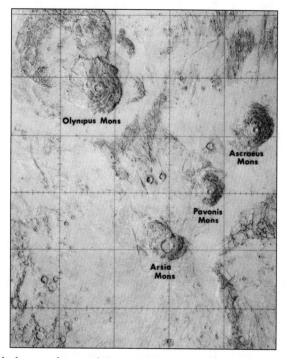

A shaded relief map of part of the northern hemisphere of Mars, containing four huge volcanoes surrounded by relatively young volcanic plains. The SNC meteorites might have formed in this area. (*U.S. Geological Survey.*)

tical about the martian meteorite proposal, and for good reason. Their calculations indicated that, while an impact might conceivably drive a golf ball–size rock off the Moon, the larger gravity of Mars would make it exceedingly difficult to blast pieces that large from its surface. Yet some SNC meteorites are even larger, as big as grapefruits, and they were probably even larger when launched because meteorites typically lose as much as half of their mass during transit through the Earth's atmosphere. The compelling evidence linking SNC meteorites to Mars nevertheless sent a few diehard researchers scurrying back to the drawing board to try to devise impact models that could accomplish this feat. Revised calculations suggest that large craters, of the size observed in the younger volcanic plains of Mars, could probably eject grapefruit- or even watermelon-size rocks from the planet. To accomplish this, underground ices might have been vaporized on impact, providing a kind of jet-propulsion assist for pieces of the target rock. A meteor that struck Mars a glancing blow and ricocheted might provide another way to propel rocks off the planet.

SOME FURTHER SUPPORT

Steadfastly resisting the idea that planetary impacts can create meteors, a few holdouts have even suggested that SNC meteorites might actually be well-traveled terrestrial rocks. In their view, these rocks were launched by impacts from Earth into great arcs or Earth-encircling "parking" orbits. Eventually, atmospheric drag slowed them down until they finally fell back to Earth, like some past man-made satellites whose gradually decaying orbits caused them to come crashing to the ground. In theory, it is possible that impact ejecta might be stored temporarily in Earth orbit. Several massive impacts on the Earth have produced *tektites*, small beads of melted rock that were sprayed over vast areas of the globe. Tektites never achieved orbit, but apparently they were heaved above the atmosphere to travel such great distances. Likening SNC meteorites to unmelted tektites is an interesting exercise, but it's dead wrong. Aside from the fact that gases trapped in shergottites are nothing at all like those in our planet's atmosphere, there is another reason that SNC meteorites cannot be from Earth. The most abundant element in these meteorites, and in rocks from anywhere for that matter, is oxygen. The oxygen now in your lungs is a mixture of three stable isotopes, atoms with different masses. The oxygen isotopes can be partly separated from one another on Earth by various processes, but always in proportion to the differences in their masses. Consequently, only certain combinations of the three isotopes are permitted, and the oxygen isotopes in all terrestrial matter, although not necessarily the same, must conform. Whether it is a rock, a fluffy cloud, tree bark, or your fingernail, everything on Earth that contains oxygen bears our planet's indelible isotopic signature. But the oxygen in SNC meteorites is distinct from Earth's, and this isotopic difference is proof that these meteorites are truly foreign.

Often in science, particularly in many geologic investigations where a definitive experiment cannot be performed, we are forced to use circumstantial evidence. In the detective saga of SNC meteorites and Mars, several bits of circumstantial evidence can be used to support this connection. For example, iron-bearing minerals in shergottites are magnetized only slightly, implying a very weak magnetic field on the parent body. Magnetization probably occurred just after impact heating, as the minerals cooled through the temperature at which magnetism is acquired. Available measurements of Mars' magnetic field by American and Soviet spacecraft are of rather poor quality, but they are good enough to suggest that a compass would probably be useless on the planet. The absence of a strong magnetic field on Mars at the present time is at least consistent with what we know about shergottites.

Chemical analyses of martian soil provide another intriguing clue. Viking landers were equipped with soil-collector arms that scooped up small shovelfuls of dirt and dumped them into hoppers in the spacecraft for analysis. Soil samples scraped from the uppermost surface were found to contain quantities of sulfur, chlorine, and bromine, which were probably exhaled from volcanoes. Terrestrial volcanoes likewise give off noxious vapors containing these elements. If we ignore these contributions to the surface soil from volcanic burps, the composition of the remaining matter is remarkably similar to shergottites. In fact, these meteorites provide the best chemical match of any known substance for martian soil. Mars dirt is not simply pulverized igneous rock, as the soil properties are best matched by an assortment of clays and oxides produced by some as-yet poorly understood weathering process. However, the observation that martian soil and shergottites have similar chemistry is one more hint that these two materials may be related.

In the quest for a home world for the SNC meteorites, almost every test imaginable has been applied at one time or another. So far, the idea that these meteorites are martian rocks has cleared the hurdles put in its path. But the only way to tell for sure will likely be to bring back a martian sample for direct comparison. Mars is a big place, and, like Earth, it is a geologically complex world. A few grab samples from one or two localities are unlikely to look exactly like shergottites, nakhlites, or chassignites, but they will carry certain diagnostic fingerprints, such as the martian oxygen isotopic composition. A Mars sample-return mission has been discussed, but its timing is uncertain and dependent on the vagaries of politics and economics. I only hope I am around to see for myself whether this idea, formulated so inauspiciously in a Texas bar, is right.

Some Suggestions for Further Reading

Bogard, D. D., and Johnson, P. 1983. "Martian Gases in an Antarctic Meteorite." *Science,* vol. 221, pp. 651–654. A technical paper describing the initial discovery of trapped martian atmosphere in shock-melted glass in a shergottite.

Dodd, R. T. 1986. *Thunderstones and Shooting Stars: The Meaning of Meteorites.* Cambridge, MA: Harvard University Press. A popular account of meteorite discoveries; Chapter 9 discusses the evidence for a martian origin of SNC meteorites.

Marvin, U. B. 1984. "A Meteorite from the Moon." *Smithsonian Contributions to the Earth Sciences,* vol. 26, pp. 96–103. A nontechnical paper that nicely describes the discovery and recognition of the first lunar meteorite.

McSween, H. Y., Jr. 1987. *Meteorites and Their Parent Planets.* New York: Cambridge University Press. This is a shameless plug for my own book on meteorites; however, Chapters 4 and 5 are relevant to martian and lunar meteorites.

McSween, H. Y., Jr., and Stolper, E. 1980. "Basaltic Meteorites." *Scientific American,* vol. 242, no. 6, pp. 54–63. A popularized description of eucrite and shergottite meteorites, and conjectures about their parent bodies.

Hardened Hearts

———

Cores and Mantles of the Terrestrial Planets

My father always had a real dislike for fancy clothes. Oh, he dressed nicely enough for work, church, and social occasions, but mostly because my mother badgered him into it. His true colors were revealed only to those who loved him. I remember one family vacation when he wore the same frayed seersucker shirt for five days running. He washed it out in the motel room sink each night and hung it up to dry, and the next morning it was miraculously ready to wear again. I believe that he positively delighted in the groans of his children at the reappearance of that faded blue-and-white striped shirt each morning. Our reactions usually precipitated a mercifully short lecture on the merits of frugality or the joys of small suitcases, which we could not escape by virtue of being imprisoned in the car with him.

My father's disdain for finery was a manifestation of a larger philosophy that he believed fervently and articulated frequently: Appearance is not an accurate guide to a person's character, or as he stated innumerable times, "Clothes don't make the man." All of us probably believe that too, although the money we spend on apparel would seem to belie our sincerity. And, unlike my father, we may sometimes make snap judgments about individuals based on superficial observations.

Like people whom we encounter only briefly, what little we know or surmise about planets is usually based on surface appearance. Planetary histories are inferred from images of towering volcanoes and yawning gorges, but their inner workings remain hidden from view. At best, we can make only intelligent guesses as to what lies inside planets. The interior of even our own world is still not fully understood. The discipline of geology is mostly devoted to studies of the Earth's crust, an exterior rind that would be thinner than an eggshell if our planet were scaled to the size of a basketball. But there are ways to peer through this shell, to spy on our planet's hardened heart.

THE EARTH'S STRUCTURE FROM SEISMOLOGY

Jules Verne, in *A Voyage to the Center of the Earth*, created an engrossing tale of descent into the Earth's interior through a volcanic crater and a series of interconnected, crystal-filled caverns. The title is misleading, as the intrepid explorers actually worked their way downward only several hundred kilometers, just a fraction of the distance to the planet's center. Most of Verne's books display amazing scientific perspicacity, but he really blew this one. We know now that the crushing pressures at even these relatively shallow depths would close any openings large enough for a worm to wriggle through. Had the adventurers somehow managed to make a journey all the way to the planet's center, they would have found conditions impossibly hostile. The pressure there is perhaps 3.5 million times atmospheric pressure, and the temperature has been estimated (with considerable uncertainty) to be as high as 6,700 degrees centigrade, hotter than the surface of the Sun. We can safely assume that humans will never have direct access to this squeezed inferno, or even want it.

A more workable, though indirect, solution to plumbing the depths of the Earth involves geophysics, a tool box of jiggles, tugging forces, and electric currents. *Seismology* is the study of how the groanings of the Earth, the vibrations generated by earthquakes or explosions, travel through the interior. Although these vibrations often are called sound waves, they have much lower frequencies than audible sound, generally being some hundred to a million times lower in pitch than the concert note A. Seismic events produce vibrations that rumble through the solid Earth along curved paths and are reflected back to the surface at certain horizons. Making sense of this cacophony of sound is a challenge that has been met only in this century. One geophysicist has suggested that reaching conclusions about the Earth's interior from seismic echoes is like trying to reconstruct the inside of a piano from the sounds it makes while crashing down a staircase. Much can be learned, however, particularly with the able assistance of supercomputers to digest the mountains of data generated by a modern global network of earthquake-monitoring devices.

A few basic rules explain how seismology works. First, the speed of a seismic wave depends on the density of the rocks through which it travels. Rock density, in turn, is related to the minerals that comprise the rock, so a seismic wave will travel faster through a rock composed of dense minerals than one containing minerals in which the atoms are not so tightly packed together. Second, any boundary separating rocks of contrasting density can either bend seismic waves so that they travel downward at a new angle or act as a mirror to reflect seismic waves back toward the surface. Third, liquids absorb some kinds of vibrations, so the seismic properties of molten and solid materials within the interior are different.

Seismology was invented as the twentieth century opened. A pioneering discovery was made in 1900 by Richard Oldham of the Geological Survey of India, when he showed that some earthquake vibrations actually travel through the body of the Earth. Then, in a 1906 paper that shook this fledgling science to its very foundations, Oldham used this insight to demonstrate the existence of the core, a world as big as Mars hidden at the Earth's center. Just a few years later Beno Gutenberg, working in Germany, analyzed seismic waves reflected from the top of the core and fixed its depth rather accurately at 2,900 kilometers. His achievement was memorialized in the name of this reflective boundary, the Gutenberg discontinuity. Another major boundary was discovered at a mere thirty-four kilometers depth by the Yugoslavian seismologist Andrija Mohorovicic in 1909. This horizon separates the crust on which we live from the underlying mantle. Like the other major seismic boundary, Mohorovicic's discontinuity was named for its discoverer; however, for reasons that should be obvious, it is now commonly referred to as the Moho. Within a span of two exciting decades and amid great rivalry among seismological laboratories, all three major structural subdivisions of our planet—crust, mantle, and core—were recognized.

On the trip to the center of the Earth, seismic waves traverse a distance of 6,370 kilometers in just a few minutes. Most seismic waves originate near the surface when a sudden slippage of fractured rocks produces vibrations that travel outward in all directions. Two kinds of seismic rumbles—compressional waves, which vibrate back and forth along the direction of travel, and shear waves, which vibrate at right angles to the direction of travel, are generated simultaneously by an earthquake. The compressional wave travels significantly faster than the shear wave. The vibrations that travel upward unleash earthquake terror on the surface. Waves propagating downward move at modest velocities through the crust and then suddenly speed up as they pass through the Moho into the denser rocks of the mantle. At about 400 kilometers depth within the mantle, seismic velocities suddenly increase again. This acceleration occurs because of pressure-induced changes in the abundant mineral olivine, a magnesium-iron silicate you might recognize as the green gemstone peridot. High-pressure experiments in laboratories have shown that the atoms in olivine respond to the crushing pressure at 400 kilometers depth by rearranging themselves into a more tightly compacted structure, called spinel. Then at 670 kilometers depth, spinel itself begins to break down and its constituent atoms are reorganized into two even

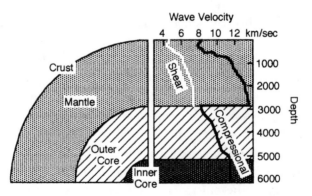

A sketch illustrating the gross structure of the Earth's interior, as well as changes in the velocities of seismic waves (compressional and shear) with depth (in kilometers). Seismic waves speed up abruptly at horizons where rock density increases.

more dense mineral structures, perovskite and cubic magnesium oxide. This change in mineralogy results in yet another jump in seismic wave velocity.

From this depth on down to the top of the core, there are no more surprises, but rather the velocities of seismic waves increase smoothly and continuously with depth. This gradual change is probably an indication that other minerals progressively transform into the perovskite structure. Perovskite dominates much of the mantle and, in fact, is probably the most abundant mineral in the Earth, yet we who live on the surface never see it. (Well, actually, this is not quite true. A rare mineral called perovskite occurs in some lavas, but its composition is calcium-titanium oxide; in the deep mantle, magnesium-iron silicate adopts the crystal structure of this mineral. Geologists have made perovskite of magnesium-iron silicate experimentally, but it survives only while held under extreme pressures in a diamond-anvil cell. You could not hold this material in your hand without it rearranging itself into another structure.)

THE EARTH'S CORE

At a depth of 2,900 kilometers, seismic waves pass through the Gutenberg discontinuity and enter the netherworld of the Earth's core. At this juncture the velocity of compressional waves suddenly plummets and shear waves disappear altogether. The explanation for this curious seismic behavior is not to be found in additional changes in the atomic packing of minerals that comprise the lower mantle. These minerals cannot be further compressed, heated, or otherwise manipulated to produce new structures that would account for the seismic properties of the core. The core must have a fundamentally different composition, but what could this mysterious matter be?

A hint was provided more than 200 years ago, when astronomer Nevil Maskelyne devised a brilliant scheme to weigh the whole Earth. Drawing on Isaac Newton's theory of the gravitational attraction between two bodies, Maskelyne measured the deflection of a plumb bob (a weight suspended from a string) by a nearby mountain, and from this observation he was able to calculate the Earth's mass. Its volume had already been deduced centuries earlier by Egyptian scholars. Mass divided by volume equals *density*, in this case the mean density of the planet. Maskelyne's calculated density was almost twice the density of rocks at the Earth's surface. It was glaringly obvious that the interior of the Earth had to contain some very dense material to account for its high mean density.

The identification of this dense matter in the core had to wait for the technology of high-pressure physics to catch up with seismology. Solids are compressible, so the velocities of seismic waves traveling through them must be measured while they are under tremendous pressures if they are to provide useful information on the core. Harvard professor Francis Birch, a genius at designing high-pressure experiments, spent years meticulously unraveling the relationships among seismic velocity, pressure, temperature, and rock composition.

From Birch's work, we now know that the Earth's massive core is made of iron, forming a huge metallic ball that undergirds the rocky mantle. Besides being dense, iron is an abundant element in the cosmos, and only a small fraction of the Earth's predicted allotment of iron can be found in its crust and mantle. A metallic iron core is a necessity if we are to reconcile the Earth's mean density with any reasonable prediction of its bulk composition. Further refinement in Birch's experiments indicates that an iron-nickel alloy actually provides a better match for core seismic velocities than does iron alone.

The absence of shear waves in the outer part of the core shows that the iron-nickel metal in this region is substantially molten. In the inner core, beginning at a depth of about 5,100 kilometers, shear waves reappear and a further jump in compressional wave velocity occurs. The inner core also is thought to be metallic iron-nickel, but here it must be solid.

THE EARTH'S MAGNETISM

It would be a mistake to think that just because the Earth's core is so remote, its influence is not felt on the surface. Its most obvious effect is seen in the swing of a compass needle, responding to the mysterious attraction of the magnetic north pole. The Earth's *magnetic field* can be visualized in a simple laboratory experiment. If a bar magnet is placed on a piece of paper that has been sprinkled with iron filings, the little bits of metal reorient themselves in a peculiar way, forming arcs that

Iron filings sprinkled over a bar magnet allow us to visualize magnetic lines of force.

loop from one end of the magnet to the other. The bits of iron allow us to see otherwise invisible magnetic lines of force. A map of the Earth's magnetic field looks basically like an enlarged version of the magnetic field generated by a bar magnet. A compass needle on the planet's surface is affected like the iron filings in our experiment, orienting itself along the planetary magnetic lines of force. The shape of the Earth's magnetic field is not exactly the same as that around a bar magnet, however, because it is distorted by ions streaming out of the Sun, as the lines of force are pushed inward toward the planet on the sunny or upwind side and swept far out into space forming a wake on the downwind side.

Certainly there is no gigantic bar magnet in the core that produces the Earth's magnetism. The interior of the planet is much too hot for any permanently magnetized material to retain its magnetic properties, so we must search for some other mechanism to account for its magnetism. In the 1950s geophysicists Walter Elsasser of Johns Hopkins University and Edward Bullard of the University of Cambridge hypothesized that stirring an electrically conducting fluid, such as molten metal in the outer core, would redirect and amplify the feeble magnetic field of the galaxy. A simple device known as a dynamo does the same thing, strengthening a magnetic field by using a rotating electric current. Exactly how such a dynamo model operates within the Earth to produce magnetism is not well understood, but the basic premise is now generally accepted.

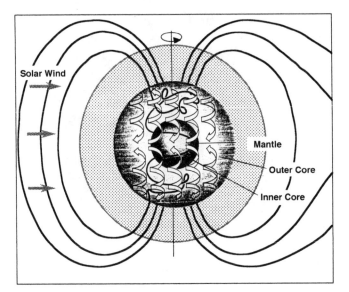

The Earth's magnetic field looks similar to that of a bar magnet, except that it is distorted by matter expelled from the Sun. Planetary magnetism is generated by movements of electrically conducting metallic liquid in the outer core. The liquid is thought to flow in screwlike rollers, through which the magnetic lines of force are threaded and twisted. (A single line of force is illustrated.)

The operation of a dynamo in the core requires that it be heated and stirred. For the Earth, these requirements may be met when heat released from crystallization of the solid inner core causes convection in the overlying liquid outer core, a stirring from below. Convection is at work in your morning cup of coffee, as hotter liquid at the bottom of the cup rises toward the surface in the center of the cup and cooler coffee sinks around the edges. In the outer core, the electrically conducting metallic liquid is thought to flow outward in screwlike rollers. But this process by itself is not enough. Magnetic lines of force must then be threaded through the rollers and twisted into pretzels. This is apparently accomplished by the Earth's rotation. The near coincidence of the Earth's magnetic and rotational poles suggests that the dynamo cannot function without reasonably rapid planetary rotation.

The Earth's dynamo has apparently been operating for millennia. Remnants of a magnetic field measured in ancient rocks point to the existence of a convecting outer core at least 2.5 billion years ago and probably 3.5 billion years ago. Thus the basic core structure must have formed fairly early in the planet's history, although the proportion of solid core has increased progressively as liquid core crystallized.

This iron meteorite, from Mount Edith, Australia, was once part of an asteroid core. Such samples provide our only opportunity to study core materials directly. The jackstraw pattern is an intergrowth of two iron-nickel alloys. The dark, round blobs are composed of iron sulfide, and the presence of sulfur in iron meteorites suggests the possibility that this is also the light element in the Earth's outer core. (*Smithsonian Museum of Natural History.*)

A LIGHT ELEMENT IN THE CORE

The view of the core as nested globes of solid and molten metal provides a rationale for its magnetism, but the idea that it is composed only of iron and nickel is an oversimplification. The outer core has a density that is slightly lower than that of molten iron-nickel, so at least this part of the core must be mixed with some other, lighter element. Which element, though, is a source of controversy. One view, not widely held but possibly correct, is that iron is alloyed with all the other elements. However, most geophysicists suspect that only one or two elements in addition to iron dominate the core.

One prime candidate for the alloying element is suggested by observations of iron meteorites, the only samples of core material that we can study directly. These metallic chunks are thought to have originally formed within cores in asteroids. Impacts subsequently demolished the asteroids, liberating core fragments that eventually collided with the Earth. Iron meteorites consist mostly of alloys of iron and nickel, but also contain a light element, sulfur, which occurs as blobs of iron sulfide. Perhaps sulfur is also the light element in the Earth's core.

Oxygen or hydrogen are also possibilities for the core's light element. At very high pressures, oxygen and hydrogen are no longer gases but, surprisingly, become metallic. Even though iron meteorites do not contain metallic alloys of iron with oxygen or hydrogen, their absence does not necessarily rule out the presence of these strange minerals in the Earth's core. Iron meteorites formed within small asteroids, where the inner pressures were too low to stabilize these metallic substances.

INTERACTIONS BETWEEN CORE AND MANTLE

Now that we have considered the composition of the core and mantle, it is worth speculating what might happen when the two touch. Experiments at high pressures and temperatures reveal that molten iron, such as comprises the outer core, reacts vigorously with perovskite, which is abundant in the lower mantle. This reaction provides a means for the core and the mantle to exchange matter. Oxygen and, to a lesser extent, magnesium and silicon from the mantle may have been added to the outer core, which possibly explains how the outer core acquired its dose of light elements. The compositional difference between the core and mantle apparently drives the reaction between them. An inverted analogy is the compositional difference between the Earth's solid crust and its liquid oceans. At this boundary, rocks and water react to produce new minerals. This analogy is limited, though, because low temperatures at the bottom of the oceans drive submarine reactions rather slowly, compared with the violent rate at which the core and mantle are thought to react.

Some scientists believe that the core-mantle boundary may be the most chemically active region of our planet. All of this flurry of thermal and chemical activity in and around the core must affect the rest of the Earth in some way. In the 1980s seismologists gained their first glimpses of the core's dynamic influence, using an exciting new technique called seismic tomography. A similar procedure, the medical CAT scan, produces a three-dimensional picture of the interior of the human body by mapping slight differences in the intensities of X rays that pass through the body in different directions. Seismic tomography generates a three-dimensional view of the Earth's interior by mapping slight differences in seismic velocities. A tomographic image illustrates where the seismic velocities are faster or slower than the average value for that depth in the Earth. Differences in seismic velocity are actually measures of density variations that, in turn, represent temperature differences, because hot material expands and so is less dense than cooler material. Tomography clearly demonstrates that heat does not rise through the mantle uniformly everywhere, but instead forms great convective plumes, employing the same self-stirring mechanism that circulates molten metal in the outer core.

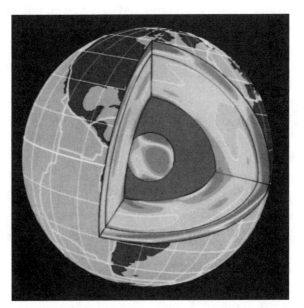

Variations in the speed of seismic waves in the mantle relate to differences in temperature. This three-dimensional, cut-away view of the Earth's interior, obtained by seismic tomography, shows hot ascending plumes and cooler regions of downwelling in the mantle. At this scale, the temperature of the outer core appears uniform, because it convects so rapidly compared to the solid mantle and inner core. *From A. M. Dziewonski and J. Woodhouse, Science 236 (1987): 37, copyright 1987 by the AAAS, with permission.*

THE INTERIORS OF THE MOON
AND TERRESTRIAL PLANETS

The information about the Earth's interior gained from this seismic journey to its center, from our consideration of its magnetism, and from various laboratory experiments that simulate conditions in the deep interior all suggest that the planet's surface appearance is a rather poor indicator of what is inside. The mantle and the core are, in many ways, distinctive worlds hidden beneath the familiar crustal globe. Can we hope to learn anything about the interiors of other planets or satellites using these techniques? Certainly we can, but our current information is very skimpy.

Seismology is the most effective tool for understanding planetary interiors, but operable vibration recorders have been installed on only one other body, the Moon. *Apollo* astronauts deployed seismometers at five lunar landing sites. Moonquake recordings, telemetered back to Earth, revealed a thick lunar crust and a layered mantle that extends almost to the Moon's center. At the very middle is a poorly defined zone that may be a small core. As useful as these data are, moonquakes are so rare and so small that the information gained about the lunar interior has been sparse. In fact, NASA elected to shut down the lunar seismic-monitoring system in 1977 rather than to pay for the continued operation of an enterprise with such modest scientific pay-off.

The *Viking* spacecraft also carried seismic recorders to Mars, but no useful information has been obtained from them. These delicate sensors were attached to the *Viking* landers' frames, and they had to be fixed tightly in place because of the stresses during launch and touchdown. The mechanism on *Viking 1* for unlocking the sensor after landing on Mars failed to respond to its activation command. The seismometer on *Viking 2* operated for a year and a half but, unfortunately, gusty winds constantly jiggled the instrument and hampered the detection of tremors. A future planned Mars mission will install a global network of seismic stations.

The study of magnetism has likewise provided only limited information. Unlike the Earth, which has magnetic lines of force looping from one pole to the other, the Moon has no magnetic field. However, rocks returned from the Moon during the *Apollo* missions record the existence of an ancient lunar magnetic field billions of years ago.

120

Drawing on what we surmise about how the Earth's magnetic field originates, the Moon may have once had a liquid core that has now cooled and solidified. Venus and Mars also have no measurable magnetic fields, and Mercury has only a small one. Because Venus is almost a twin of the Earth, it seems likely that it too should have a molten outer core. But Venus rotates very slowly, only once in 243 days, and this rate may be too slow to slosh the metallic liquid in its core and create a dynamo. Mercury's slow rotation, once in 59 days, also may inhibit magnetism, even though it apparently has a massive core. The absence of a martian magnetic field is puzzling, because Mars rotates at about the same rate as the Earth. Perhaps its core is solid throughout.

Without the benefit of seismic recordings, and with very little useful information from magnetism, how much can we really hope to understand about the interiors of other planets? At present, planetary geophysicists are bedeviled by lack of data, and so must tackle the problem with theory. The approach involves making imaginary models of the planet's interior that do not contradict any information that is known about the planet. Several kinds of observations, which we can make for any planet without actually placing instruments on its surface, are used as constraints for the model-building exercise.

The mean density of the planet is one such piece of information. To derive this we require, like Maskylene, a plumb bob to calculate the planet's mass, or at least another object that feels its gravitational pull and responds in some measurable way. A spacecraft makes an excellent plumb bob, because its trajectory is deflected by the nearby planet. Nowadays, a planet's dimensions can be measured in telescopic images or, more precisely, from the loss of spacecraft transmissions in its shadow. Dividing the mass by the volume gives mean density.

Trying to make conclusions about planet compositions based on mean densities would cause chaos if we were free to choose from the entire list of elements in making planet models, but, luckily, the list can be shortened considerably. We know that none of the terrestrial planets are made primarily of obscure elements such as tin, neodymium, or krypton; instead, they must be composed of the same dominant elements as in the Earth—silicon, oxygen, magnesium, iron, and a handful of others in lesser quantities. When only these common elements are considered, the number of permissible models is reduced significantly, so that mean density variations can be understood in terms of fundamental differences such as the amount of metallic core versus rocky mantle.

Another constraint on planet models is the way in which materials of different density are arranged inside the planet. We can rule out some models for planetary interiors immediately. For example, no one would now argue that planets are hollow, even though reputable science fiction writers in the past adopted this idea. Edgar Rice Burroughs' *At the Earth's Core* described the underworld of Pellucidar populated with dinosaurs, and William Reed's *Phantom of the Poles* featured oceans and continents on the inner surface of a hollow Earth that could be reached by sailing through openings at the poles. The idea of hollow planets may seem naive in light of modern geophysics, but it is not necessarily obvious that other terrestrial planets must have Earthlike interiors with very dense matter at their centers. We gain insights into how materials of different density are distributed within a planet from the way the body responds to rotation. A rapidly spinning planet, such as the Earth, is not spherical—it bulges at the equator and is flattened at the poles, and the size of the bulges is related to internal density variations. Slowly rotating bodies, such as Mercury or Venus, are not flattened measurably, however, so internal density distribution cannot always be determined.

Construction of a planetary model begins with an estimated bulk composition. The variation in pressure with depth is then calculated, and minerals known to be stable at different pressures are assigned to various depths. Of course, once the model is completed, the chemical compositions of all the minerals must sum to the correct bulk chemical composition of the planet. Next, geophysical traits, such as mean density, of this particular planet model are calculated and compared with the actual values measured for that planet. If they do not agree, the model must be either adjusted or abandoned. Naturally, there is no unique solution in this kind of planetary modeling, but models can be tested rather stringently if enough constraints are available.

Acceptable, though possibly inaccurate, models of the terrestrial planets show significant differences in the proportions of rocky mantle and metallic core. Mercury is an unusual planet that is mostly core. The interior of Venus is rather Earthlike, at least to our current limited state of knowledge. Mars appears to have a small core, possibly made of iron sulfide, and the planet's small size precludes many pressure-

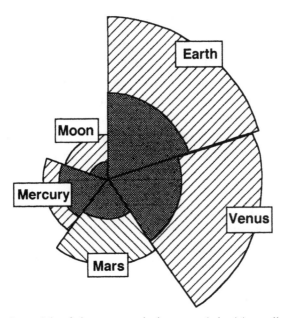

Slices through models of the terrestrial planets and the Moon illustrate differences in the proportions of massive central cores and rocky mantles.

induced changes in the mineralogy of its mantle. The Moon has a very thick crust and mantle and a tiny core, perhaps made chiefly of iron sulfide. As you might expect, the lunar model is better constrained than those for other planets, because we have at least limited seismic data and a fair estimate of its bulk composition based on returned lunar rocks.

Although models of planetary interiors point to some fascinating conclusions, we should keep in mind that they are a lot like value judgments of people based on how white their teeth are and what kind of car they drive. We have very little hard information with which to test planetary models, and some of the assumptions made during their construction may be wrong. And new revelations about the Earth's interior should give us some pause in believing that we actually know a lot about the insides of other bodies. The Earth is a reasonably well instrumented geophysical laboratory, and if we are still being surprised by the bizarre and dynamic nature of its interior, think what secrets must be buried in the hardened hearts of other planets.

Some Suggestions for Further Reading

Allegre, C. 1992. *From Stone to Star*. Cambridge, MA: Harvard University Press. A delightful, general-audience book on the evolution of modern geology; Chapter 2 gives a good overview of the historical development of our understanding of the Earth's interior.

Anderson, D. L., and Dziewonski, A. 1984. "Seismic Tomography." *Scientific American*, vol. 251, no. 4, pp. 60–68. This article describes the exciting new picture of the Earth's interior gained from seismic tomography.

Bolt, B. A. 1982. *Inside the Earth: Evidence from Earthquakes*. San Francisco: Freeman. Everything you ever wanted to know about seismology, written in an entertaining and understandable way; Chapter 4 includes sections on seismicity of the Moon and Mars.

Bott, M. H. P. 1982. *The Interior of the Earth: Its Structure, Constitution and Evolution*, 2nd ed. New York: Elsevier Science Publishing Company. An excellent textbook on the nature of the Earth's interior; a technical but nonmathematical treatment of geophysics.

The Dynamic Earth. 1983. New York: Scientific American. A series of interesting articles by authorities in the field, discussing various aspects of Earth science. Chapters by R. Jeanloz, "The Earth's Core," and by D. P. McKenzie, "The Earth's Mantle," are especially relevant to the topic of the Earth's interior.

Jeanloz, R. 1990. "The Nature of the Earth's Core." *Annual Review of Earth and Planetary Sciences*, vol. 18, pp. 357–386. An authoritative account of current thinking and research on the nature and composition of the core.

No Stone Left Unturned

Regoliths on the Moon, Asteroids, and Planets

One of my great pleasures, one might even say consuming passions, is eating homegrown tomatoes. For me, nothing rivals the luscious, acidic taste of a vine-ripened, freshly picked tomato. Each spring I laboriously cultivate a small garden of tomato plants and watch impatiently for the appearance of the yellow flowers, the tiny green fruit, and finally the first mature red tomato. Then, within a few short weeks, my dozen plants rapidly begin to produce ripe tomatoes faster than I can eat them, a bounty (or glut, according to my wife and daughter) that requires a hedonistic binge of consumption.

What makes this annual tomato frenzy possible is good dirt. My backyard, however, is not blessed with this commodity, so I must import it. Each year I acquire a truckload of what the local residents call mushroom dirt, a mixture of compost and mulch that has been discarded by a nearby mushroom-growing concern after producing a crop of their edible fungi. My tomato garden is composed of approximately equal parts of this fragrant material and the native Tennessee soil, all thoroughly mixed with the aid of a tiller.

Even though there are no tomato connoisseurs on the Moon, the surface there too has been gardened, ground into rock powder and periodically overturned. The layer of loose rock fragments and dust draped over the lunar surface is called the regolith, from Greek words meaning "stony blanket." Unlike soil on the Earth, which is produced primarily by interactions among bedrock, the atmosphere, and living organisms, the Moon's regolith formed through countless meteor impacts. Millennia in, millennia out, continuous bombardment by small meteors has pulverized the surface rocks. Larger impacts intermittently churned this powder, as well as excavated blocks from the bedrock below and mixed them with finer particles. The result of this cosmic gardening is a monotonous gray blanket of unconsolidated rubble of varying sizes, ranging from large boulders to tiny flecks of dust.

EARLY IDEAS ABOUT THE LUNAR REGOLITH

The fact that the Moon is covered with a regolith has been known for some time, since before the *Apollo* era and even the earliest close-up photographs of the lunar surface by unmanned spacecraft. When you look at the full Moon, you see that its surface is uniformly bright from one limb to the other, even though the intensity of light reflected from a smooth sphere should diminish toward its edges. From this observation, astronomers in the middle of this century deduced that the lunar surface had to have the optical properties of dust. But is this powder just a thin surface coating, or does it extend to great depth?

As you might imagine, this question was of considerable concern for those involved in planning for *Apollo* astronauts to land on the Moon. Although most scientists at the time argued for a coherent layer of dust perhaps a few meters thick, some suggested the plausibility of a kilometer or more of fine regolith material. They feared that such a thick dust layer would not support the weight of an *Apollo* landing module. The question of whether man could actually land and walk around on the Moon had to be answered by actually going there, what geologists sometimes call "ground truth."

In this case, ground truth came in the form of an ungainly, man-made but unmanned object called *Surveyor I*, the first spacecraft to soft land on the Moon. Several *Ranger* (American) and *Luna* (Soviet) spacecraft had "hard-landed," which is to say crashed, earlier on the lunar surface, but when the first *Surveyor* plopped down gently in 1966 our perspective of the Moon changed forever. *Surveyor's* soft landing was accomplished through a programmed sequence of events orchestrated by an onboard computer, which at the time was something of a technological triumph. As the craft neared the Moon, retro-rockets slowed it down from nearly 6,000 to several hundred miles per hour, at which point a Doppler radar system began measuring its height above the lunar surface. The designers correctly gambled that the radar signal would be reflected from the top of the regolith rather than from rock buried beneath nonreflecting dust. The radar system, sensing its looming target, activated rockets that further bled off speed until, at a height of just twelve feet above the surface, *Surveyor* was descending at a mere three miles per hour. Then the rockets shut off and the spacecraft simply dropped the last twelve feet, making a perfect three-point landing. Quickly the engineers in charge of the mission sent a signal that pointed the onboard camera at *Surveyor's* feet and took a picture. An enthusiastic cheer from hundreds of onlookers at Caltech's Jet Propulsion Laboratory greeted the arrival of this photograph went it was received on Earth several minutes later. *Surveyor's* landing pads were resting firmly on top of the lunar regolith. The thump of arrival had created shallow depressions about an inch deep in which each landing pad sat and had kicked up a little dirt, but all three pads were fully visible. The weight resting on each pad was 33 pounds—approximately 100 pounds total for the whole spacecraft—although its total weight back on Earth had been 600 pounds due to the greater gravitational force there. From this photograph, it was immediately clear that astronauts could walk on the Moon without sinking out of sight, and the idea that the regolith could not support the weight of a landing module bit the dust.

NATURE OF THE LUNAR REGOLITH

We now know a lot more about the Moon's crumbled covering, mostly because of revelations made during the *Apollo* program. The most striking characteristic of the regolith is the completeness with which it covers the lunar landscape. This blanket produces a subdued topography everywhere, smoothing the angular prominences and craters on the Moon's surface. It is not intuitively obvious why a flattened

An astronaut's view of the lunar regolith, a pervasive blanket of rocky blocks and powder pulverized by countless impacts. This photograph was made at the *Apollo 15* site. *(NASA Johnson Space Center.)*

surface should be formed by firing the equivalent of meteoritic bullets at it, a problem that has worried students of the Moon for some time. The explanation lies in the fact that larger crater depressions have been filled by the ejecta from superimposed smaller craters, in an endless exercise of overlapping circles that only patient Mother Nature could manage.

Most vistas of the lunar regolith appear lumpy, with large blocks embedded in finer material. Such large boulders were the source of the Moon rocks returned by the *Apollo* astronauts. Bedrock is completely obscured by regolith in most places and was never sampled during any of these missions. If large rocks in the regolith were targets of opportunity for the astronauts, the fine dust was a source of aggravation, as it formed an ever-present coating on their spacesuits, instruments, and lunar rovers. This gray, gritty flour also provided a template on which to record evidence of the astronauts' presence, a substrate on which their boot pawings and tire tracks were imprinted as they scurried about on their assigned tasks. Based on the estimated rate of regolith turnover, these markings will probably last for thousands of years before they are finally erased by small impacts. Of all the wonderful photographs of the *Apollo* era, my favorite is the famous close-up picture of Neil Armstrong's bootprint in the lunar soil, a kind of fossil that eloquently proves that man was here.

The thickness of the regolith, of such concern to *Apollo* mission planners, turns out to vary from place to place, depending on the age of the surface on which it formed. The average regolith thickness on the maria is about five meters, only about half the depth developed on older highlands terrains. Even though we have thickness estimates from a number of locations, specifying exactly where the bottom of the regolith is turns out to be a difficult job. The rock underlying the loose, unconsolidated regolith is itself badly fractured by impacts. This smashed layer, extending to depths of several kilometers, forms a transition zone between the surface regolith and the undisturbed rock below.

The rate at which impacts reduce rocks to rubble on the lunar surface is apparently very slow, requiring many millions of years to create enough soil to grow a tomato plant. In that same amount of time on the Earth, whole mountain ranges are routinely worn down to nubs, and incredible sedimentary piles many kilometers thick accumulate in depressions. Of course, we have cascading streams, glaciers, temperature fluctuations, windstorms, burrowing animals, and plant roots to thank for this. Such efficient weathering and erosion cannot be matched on the dry, lifeless Moon. Our planet, like the Moon, has experienced the tortoiselike cumulative effects of meteor bombardment, but rock destruction by this method is swamped by the much more rapid soil-producing mechanisms of a wet, living world. From a scientific perspective, though, the ponderous development of the lunar regolith offers a distinct advantage: This slowly accumulating blanket of rubble contains a historical record of billions of years of lunar surface processes and, as we will see shortly, the Sun's evolution.

The regolith record can be deciphered only by examining the details of its structure from returned lunar samples. Two different kinds of regolith samples were collected by the *Apollo* astronauts. The first was a "grab" sample taken by scooping or raking loose soil into a Teflon bag. On each mission, the first astronaut to the ground immediately collected at least one grab sample from the site as a contingency in case some emergency forced a rapid evacuation. The second kind of regolith sample was a core, which was collected inside a long aluminum barrel that was screwed or hammered into the ground. This "drive tube" was fitted on one end with a disposable steel bit and on the other with a handle; the bit was replaced after use with a cap to retain the core sample.

The longest drive tube samples were about ten meters in length and had to be broken into sections for the trip back to Earth. Unlike grab samples, these cores preserved the fragile stratigraphic relationships within the regolith. Some cores were homogeneous, because the materials in them were churned by a single meteor impact. Others, however, had subtle but discernible layers. For example, one core from the *Apollo 15* site contained forty-two separate stratigraphic units, some no thicker than a pie crust. The principle of superposition tells us that the layers at the bottom of a drive tube are older than those near the top. However, the sequences of layers in adjacent core tubes often cannot be matched, so the impacts that produced them must have been very localized. The thin layers in regolith cores represent

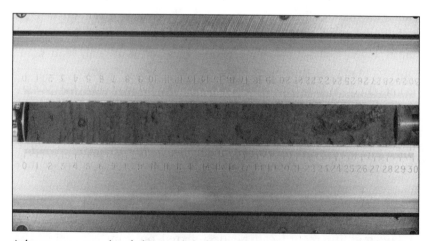

A lunar core sample of the regolith from the *Apollo 15* landing site. The fine laminations visible in the core were former surfaces of the Moon, each buried by ejecta from another small impact. (*NASA Johnson Space Center.*)

a succession of lunar surfaces, each a thin blanket of impact ejecta that was subsequently overlain by another, newer surface.

By dissecting the cores and studying the individual particles under a microscope, we can learn something about the sources of the material that were pulverized to form the regolith. Because the composition and texture of soil vary from site to site, geologists have inferred that the lunar regolith was mostly derived locally from the underlying bedrock. For example, regolith samples taken from sites underlain by the dark maria mostly contain tiny fragments of congealed mare lava. However, at each site some small fraction of the particles is clearly exotic, lofted great distances by very large impacts. At the *Apollo 11* site, about 5 percent of the regolith is thought to have been derived from as far as 100 kilometers away, and half a percent may have traveled more than 1,000 kilometers.

If lunar soils were made by pummeling lunar rocks, you might well ask where are the meteoritic projectiles that accomplished all this. It seems reasonable to assume that they were broken up and mixed into the regolith. Systematic examinations of soil samples, however, have turned up precious few recognizable bits of meteorite. My own first published meteorite paper was on this very topic. That article described a millimeter-size chunk of meteorite found in soil that was collected at the *Apollo 12* site. This unusual fragment was different from meteorites that fall to the Earth today, probably because it was part of a population of objects that orbited at a much earlier period in solar system history. A handful of tiny meteorite fragments have been recognized by other researchers studying lunar soils, but finding one is an uncommon event and requires some luck. The question, then, still remains: Where are these meteoritic projectiles? The answer is that the force of impact pulverized them, so that all that remains are dust particles too tiny to recognize as meteorites. We know that their minute, shattered remains are there in the lunar soil, however, because they imparted a distinctive chemical signature to it. By analyzing the soil for elements such as nickel, gold, and iridium that occur in much higher proportions in meteorites than in Moon rocks, researchers have documented the nearly invisible meteoritic component in the lunar regolith.

A curious property of lunar soil is that it ages, or "matures," as its residence time on or near the surface increases. The most obvious characteristics of soil maturity are smaller grain sizes and increased amounts of glass. In immature soils, the individual particles are fragments of rocks or minerals of varying sizes, accompanied by a few little droplets of glass. The glass particles are of course melted rocks, splashed by impacts and quickly cooled before they had a chance to crystallize. The hot melt commonly adheres to the crystalline fragments and glues them together into delicately welded clumps. Mature soils are more completely crushed into finer fragments and have higher proportions of glass.

REGOLITHS ON OTHER BODIES

The Moon is the only extraterrestrial body we have studied with any degree of thoroughness, so its regolith forms the basis for our understanding of how such surface materials form. But all the terrestrial planets have regoliths of some sort, although we know very little about them. *Mariner 10*'s photographs of Mercury were taken from too far away to provide information on its regolith, but it is probably similar in many respects to lunar soil. Photographs of the surfaces on Mars and Venus clearly show that loose dirt also occurs on these planets. Both of these planets have atmospheres, so their regoliths probably did not form by meteor bombardment alone. The formation of venusian soil may be assisted by the chemical weathering of rocks in contact with its dense, corrosive atmosphere, and martian soil-forming reactions may be driven partly by the Sun's ultraviolet radiation. Only on dry, airless bodies such as the Moon should we expect to find regoliths generated solely by impacts.

In our solar system, there is no shortage of dry, airless bodies, but most of them are tiny, perhaps a few tens of kilometers in diameter. Materials dislodged by impacts at speeds no greater than a softball toss can readily escape from these small bodies, so it is not obvious that they can retain regoliths. Thousands of such objects occupy orbits in the asteroid belt between Mars and Jupiter. The *Galileo* spacecraft dashed by asteroid Gaspra in 1991. Gaspra is an oblong boulder some twenty kilometers in longest dimension, pockmarked with numerous craters. The resolution of Gaspra's image is such that we cannot tell whether it has a regolith, but the subdued appearance of its craters may suggest a thin covering of dust.

Panoramic views of the surface of Mars (by *Viking 1*, above) and Venus (by *Venera 13*, below) show regoliths. Mars has sand and abundant angular blocks. In contrast, the Venus site has exposures of bare rock partly covered by unconsolidated soil. (*Jet Propulsion Laboratory.*)

The tiny moonlets of Mars, Phobos (twenty-seven kilometers in longest dimension) and Deimos (fifteen kilometers long), are of asteroid size, and were probably captured from the adjacent asteroid belt. High-resolution photographs of the surfaces of both satellites, obtained by Viking orbiters, clearly show that they are thickly blanketed with dust. However, the formation of regoliths on these moons might be due partly to the fact that they are in orbit around Mars, so that any material ejected from their surfaces also would orbit the planet and have the opportunity to be swept up again.

If bodies in the asteroid belt have regoliths, it seems likely that some unconsolidated regolith material would be dislodged and find

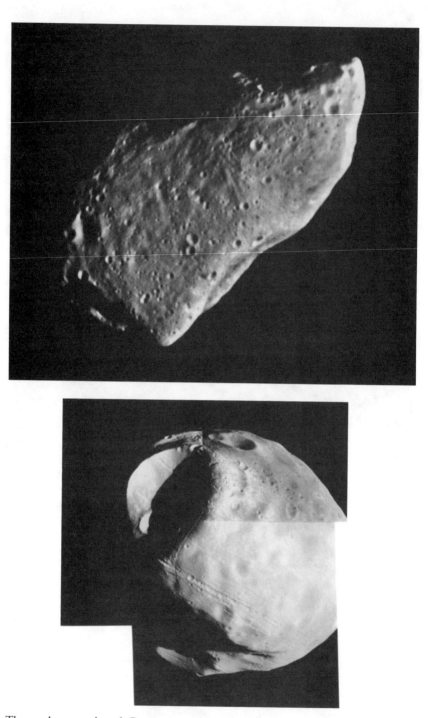

These photographs of Gaspra, a small asteroid (above), and Phobos, a tiny moon of Mars (below), feature subdued craters that may be partly filled with dust and rubble. Their appearance suggests that asteroids are probably covered with regoliths. *(Jet Propulsion Laboratory.)*

This meteorite, from Dwaleni, Swaziland, is a breccia composed of angular rock fragments cemented together with finely ground dust. This regolith sample from an asteroid is similar to soil breccias found on the Moon. (*Smithsonian Museum of Natural History.*)

its way to Earth. Such dust motes probably would not survive transit through the atmosphere. However, some hardened clumps of asteroid soil, called regolith breccias, have arrived as meteorites on the Earth. These objects provide incontrovertible evidence that some asteroids can retain regoliths.

Regolith materials on the Moon have been compacted and cemented together in a similar way to form soil breccias. Compared to lunar soil breccias, however, meteorite regolith breccias have larger grains and very little glass. In other words, asteroid regoliths are less mature than that on the Moon. There are several possible reasons for this difference. The asteroid belt contains many orbiting objects, so traffic accidents should be more frequent in this region than in the vicinity of the Moon. As a consequence, the rates at which rocks are pulverized and the resulting dust is gardened may be faster on asteroids. At first glance, this train of logic might seem to argue for more mature regoliths on asteroids than on the Moon. However, more impacts would also lead to rapid burial of asteroid surfaces so that they would become protected from later smaller impacts. Asteroids also travel slower than bodies that impact the Moon, leading to less destructive collisions. Moreover, the weaker gravity of asteroids, relative to the Moon, would allow ejecta to travel farther, probably dusting the whole body rather than just an area adjacent to the source crater. The bottom line is that many asteroids may be shrouded with thick dust mantles. The fears of *Apollo* planners about the thickness of the lunar regolith proved to be groundless, but they might be realized on asteroids.

A RECORD OF SOLAR ACTIVITY

Back on Earth, my tomato garden flourishes because of good dirt, but also because these plants bask in sunlight. Even in the absence of growing plants on the Moon, the effects of the Sun's rays can be seen in its regolith materials. Visible light is only one of the forms of radiation that the Sun emits. Others such as ultraviolet radiation are less obvious to our limited human senses but, as every sunbather knows, are no less powerful. One type of radiation that is harder to detect is the *solar wind*, which consists of ionized gases, mostly protons and electrons, heated to 100,000 degrees centigrade. This wind is actually an extension of the corona, the outer envelope of the Sun, comprising boiled-off bits and pieces of stellar matter that the Sun's gravity cannot hold in place. The solar wind steadily streams outward from the Sun's corona at a speed of 400 kilometers per second until it finally dissipates in the emptiness of space far beyond the confines of the planetary region. We do not know the exact point where it merges with the interstellar medium, but there is hope that a distant *Voyager* or *Pioneer* spacecraft will eventually reach this point and radio back its location to us. The solar wind is an exceedingly dilute gas, a gossamer breeze that cannot be felt at all by humans. At the Earth's orbital distance, the density of the solar wind is typically five particles per cubic centimeter under quiet solar conditions, but sporadic fluctuations in the Sun's activity can produce gusts of solar wind. As these particles approach the Earth, they are deflected by the planetary magnetic field or are imprisoned within it, forming concentrations of ions such as the Van Allen radiation belts.

The Earth's atmosphere provides a protective shield that screens out most solar wind particles (luckily for us, because solar wind irradiation can cause genetic mutations). However, the Moon and other airless bodies offer no refuge from the Sun's relentless breeze. While this situation may be quite rightly a subject of some concern for future populations that might someday inhabit the Moon, it is also scientifically useful. The exposed lunar regolith contains a record of the Sun's evolutionary history, written in expelled solar wind. The regolith is, in this sense, like a photographic film that has captured the Sun's likeness in hues of nonvisible radiation.

Solar wind particles cannot penetrate very far into mineral grains in the lunar regolith, but a great many of them bore in persistently on the target. The mineral grains in lunar soil are commonly decorated with numerous solar wind pits, usually thousands within an area the size of a fingernail. Although solar wind particles produce surface damage on mineral grains, they can punch themselves only a minute distance into the target grains before coming to rest. Their paths through the crystals are marked by straight tracks, like tiny bullet holes riddling a target. Because these particles do not have enough energy to penetrate below the very surface of the regolith, they are dependent on gardening by small meteors to turn the soil and roll the grains around so that each side is exposed to the shower of radiation from above. Judging from the uniformity of track distribution in most regolith grains, the lunar soil has been turned constantly, like a sunbather who carefully times exposure on front and back.

Just as I plant tomato seeds in my garden, the solar wind particles plant themselves inside target mineral grains, though more violently. The lunar regolith contains much higher amounts of hydrogen, helium, carbon, nitrogen, and sulfur than do the rocks from which they formed, a legacy of solar wind implantation. In fact, virtually all of the hydrogen atoms in lunar soil are from this source. The isotopic composition of solar wind gases in lunar soils tells an interesting story. The ratios of helium isotopes and of nitrogen isotopes in the regolith are different from the present-day solar wind, implying that the proportions of these isotopes must have varied over the lifetime of the Sun. Although we do not yet fully understand the reasons for these changes, they must be tied to the Sun's thermonuclear evolution.

In the army of irradiation, solar wind particles are merely shock troops, softening the target for attack by the more violent *cosmic rays*. These particles, zooming in from outside the solar system, are more energetic and can bore several meters into the regolith. As these tiny projectiles smack into other atoms in the target, they dislodge protons and neutrons and thereby transform the atoms into new elements. Scientists put this alchemy to good use to measure the length of time the regolith has been exposed to cosmic rays. Because even energetic cosmic rays cannot penetrate too deeply into the soil profile, the time of cosmic ray exposure represents the duration of a regolith sample's residence on or very near the surface. Residence times on the lunar surface of a few million years or so are common.

Although irradiation of the regolith on the airless Moon or an asteroid is more extensive than on the Earth, the amount of particle impacts is small during, say, a typical growing season. Therefore, this should not provide any serious impediment to growing tomatoes on one of these bodies. You may have read newspaper accounts of the millions of tomato seeds that were carried aloft in the *Long Duration Exposure Facility*, a satellite that was stranded in Earth orbit for six years following the *Challenger* disaster. These seeds received radiation doses 5,000 times as great as their Earthbound counterparts. After the satellite was finally retrieved by a space shuttle, the seeds were distributed to millions of schoolchildren to plant and observe. The idea behind this modest experiment was to gauge the possible effects of an increased amount of cosmic irradiation on tomato genetic matter. Most of the seeds germinated and the tomatoes grown from them were normal, although there was some unfounded concern expressed about whether they were safe to eat. My point is, though, that tomato seeds apparently can survive cosmic irradiation. All that may be needed to grow tomatoes on the Moon is some carbon dioxide to facilitate photosynthesis, a little water, a moderated temperature, and some dirt. And the dirt is already there, tilled and ready for planting. This soil may need a little fertilizer, but there seems to be no reason why it could not be used for farming if enclosed within a suitable atmospheric dome. Life for future moonlings will be hard in many ways, but I find great comfort in the knowledge that they should not have to live without homegrown tomatoes.

Some Suggestions for Further Reading

Beatty, K., O'Leary, B., and Chaikin, A., eds. 1982. *The New Solar System*. Cambridge, MA: Sky Publishing Corporation. A wonderful book that captures the excitement of solar system research; Chapter 2 by J. A. Eaddy describes the solar wind and Chapter 7 by B. M. French discusses the lunar regolith.

French, B. M. 1977. *The Moon Book*. Baltimore, MD: Penguin Books. A richly illustrated, nontechnical review of lunar science, including studies of the regolith.

Langevin, Y., and Arnold, J. R. 1977. "The Evolution of the Lunar Regolith." *Annual Review of Earth and Planetary Sciences*, vol. 5, pp. 449–489. An excellent but rather technical review of the state of knowledge of the lunar regolith.

McSween, H. Y., Jr. 1987. *Meteorites and Their Parent Planets*. New York: Cambridge University Press. A book on meteorites for the nonspecialist; meteorite regolith breccias are described in Chapter 3 and cosmic-ray exposure is explained in Chapter 8. If you don't like the book you're reading now, you won't like this one either.

Taylor, S. R. 1982. *Planetary Science: A Lunar Perspective*. Houston, TX: Lunar and Planetary Institute. This is a marvelous book detailing what has been learned about lunar geology; Chapter 4 describes the lunar regolith in understandable terms.

And Not a Drop
to Drink

———

Water on Mars

I once flew across the central United States in a small airplane, at an altitude of just a few thousand feet. This trip was an eye-opener, not at all like soaring above the clouds in an airliner at thirty or forty thousand feet. What I remember most was a seemingly endless expanse of large, verdant circles, the products of center-pivot irrigation systems used by Kansas and Nebraska farmers to water their crops. The geometric monotony of these perfectly laid out fields was broken only by an occasional meandering river. What was most noticeable in viewing this prairie land from a mile above was water, or at least its erosive and cultivating consequences.

Landsat photographs, taken from a satellite platform orbiting many hundreds of miles above the surface, show the same features that caught my eye during this flight at low altitude. The Euclidean geometries of the man-made circles are much more apparent than the rivers, however, because they stand in such stark contrast to nature's more random constructions. Of course, the resolution of this patchwork quilt is lost rapidly as viewing distance increases. It is a fact, nonetheless, that some of the first faint indications of intelligent life on Earth, as viewed from great distance, are provided by the patterns of our irrigation systems.

EARLY VIEWS OF WATER ON MARS

More than a century ago, the Italian astronomer Giovanni Schiaparelli reported sightings of a network of straight lines crisscrossing the surface of the planet Mars. Schiaparelli described them as grooves, naturally using his native language. The Italian word for grooves is *canali*. Unfortunately, this term was translated into English as "canals," a word that connotes design and construction by intelligent beings. Although Schiaparelli never suggested that his grooves might be filled with water, the world was shortly awash with imaginative tales of the imposing system of irrigation ditches built by martians. H. G. Wells' *War of the Worlds*, a science fiction classic describing a martian invasion of Earth, was published just two years after Schiaparelli's report and was undoubtedly influenced by this scientific bombshell.

When Schiaparelli announced in 1892 that he was giving up observations of Mars because of failing eyesight, Percival Lowell took his place as the world's most celebrated Mars-watcher. Lowell was a Bostonian with a penchant for planets and a hefty pocketbook that allowed him to pursue his astronomical interests. With his own funds, Lowell constructed an observatory in Flagstaff, Arizona, where he spent long nights carefully sketching the martian surface features he spied through his telescope.

Lowell believed that he saw incontrovertible evidence of an intelligent race inhabiting Mars. His martians lived on a dessicated world, much like the Arizona desert that housed his observatory, but they coped by constructing gigantic systems of canals to carry water from the polar caps to the thirsty populations in equatorial cities. Lowell noticed that the water from the martian canals also made the arid desert bloom, as evidenced by his observations of seasonal changes in the dark areas of the planet's surface.

The exotic view of martian civilization that Lowell conjured up was immensely popular with nonscientists, despite the fact that a number of astronomers had failed in their own attempts to confirm the observations of canals that were the backbone for his speculations. Contemporary scientific skepticism about Lowell's ideas was perhaps best personified in Alfred Wallace, a respected researcher who had independently proposed the idea of evolution by natural selection (although Charles Darwin has since gotten all the credit). As Wallace had been educated as an engineer rather than as an astronomer, his arguments against Lowell's martian civilization were based on an assessment of the planet's habitability rather than the reality of canal sightings. In a scathing review of one of Lowell's books, he revealed that Lowell had made a mistake in estimating the average surface temperature on Mars. Rather than being comfortably cool like "the South of England," martian temperatures everywhere were below the freezing point of water. Moreover, the martian atmosphere was thinner than Lowell had calculated, with the result that any liquid water on the surface would rapidly evaporate if it did not freeze. Wallace concluded that Lowell's evidence for intelligent beings on Mars, drawn primarily from his notions about water on the planet's surface, was fantasy.

We do not know what Percival Lowell really saw in the blurry martian images glimpsed through his telescope, but it certainly was not canals. On closer inspection from orbiting spacecraft, not a single linear feature has been seen at the locations where Lowell sketched his sluices. And Wallace's harsh analysis of martian hydrology has been confirmed. Mars has no lakes or streams, and the lifetime of any liquid water on the planet's surface can be measured in minutes. All this is not to say, however, that there has never been any water on Mars in the past.

MODERN SPACECRAFT OBSERVATIONS

By the past, I certainly do not mean a century ago, when Lowell made his observations. Water last ran downhill on Mars at least a billion years ago, or perhaps several billion years ago. But run it did, in trickles and streams, and in cascades and torrents. The telltale evidence for this running water comes from two distinctive kinds of gouges in the martian surface, first photographed in 1971 by the *Mariner 9* spacecraft and later in 1976 by *Viking* orbiters.

The first of these remarkable features are *outwash channels*, a relatively bland term that does little justice to their violent origin. These huge gouges, hundreds of kilometers long, were formed by gully-washers, catastrophic floods on a scale unknown on the Earth. Outwash channels are concentrated near the martian equator, where they

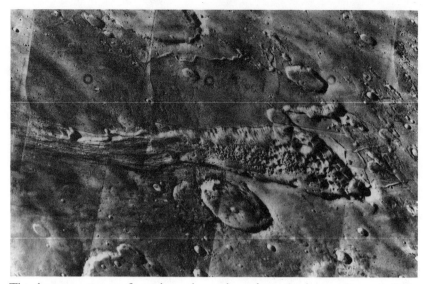

This huge martian outflow channel may have formed when underground ice melted, causing sudden collapse of the depressed, chaotic terrain on the right, and flowed downhill in torrents. The downstream end of the channel, at the left, fades into obscurity, as if the water just evaporated. (*Jet Propulsion Laboratory.*)

arose within the southern highlands and flowed out onto the lower plains to the north. Their source areas are depressions filled with jumbled, tilted blocks. The foundation under this chaotic terrain must have been suddenly mined away, leaving the unsupported roof to collapse like a ton of bricks.

The outwash channels on Mars are similar to the geologic ravages caused by catastrophic floods on the Earth, but the martian versions are much larger. As much as half a billion cubic meters of water per second, a thousand times the discharge of the Mississippi River at its mouth, may have gushed through these sluices. These torrential floods, carrying suspended sediments and perhaps blocks of ice, scoured and sculpted the martian surface before them in their relentless rush to lower elevation. They left behind twisted mazes of deep gullys, and hills isolated and molded into streamlined, teardrop-shape islands as the water surged around them. As the floods subsided, they released their suspended sediments, forming sandbars and braided deposits along their length. The valleys at the downstream ends of these channels often simply fade into obscurity. It is as if all this water just disappeared into thin air, which is exactly what may have happened.

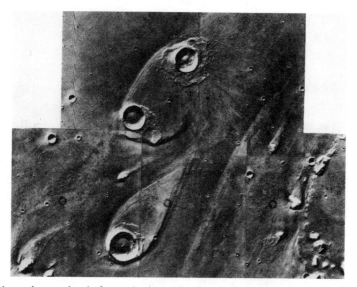

Teardrop-shape islands formed where the flood was diverted around the walls of large impact craters in its path. *(Jet Propulsion Laboratory.)*

The ages of the outwash channels have been estimated from the abundance of superimposed impact craters. There is considerable uncertainty in these ages, however, because we have not calibrated the rate at which meteors pelted Mars over time, as has been done for the Moon. Ages of the channels estimated from crater counts range from 1 billion to 4 billion years old, suggesting that they formed sporadically over a long period of time.

How can such channels exist if water cannot survive on the martian surface? As you might expect, this conundrum has caused considerable controversy. A few geologists have even suggested that running water did not cut the channels, but instead they formed through erosion by lava or glacial ice, wind action, or floods of petroleum. There is little evidence for these assertions, though, and the clear consensus is that these incised valleys are the work of water. Where, then, could this torrent of liquid water have come from? As water cannot persist on the surface of Mars, the inescapable conclusion is that it must have been stored underground, presumably under the chaotic terrains that lie at the headwaters of the channels. Perhaps these areas contained underground ice, called permafrost. If so, some source of heat, such as molten lavas ascending from below, defrosted the frozen ground so that water escaped and carved the channels. This is an attractive explanation, but the calculated volumes of ice removed from the source terrains are too small in some cases to account for the huge floods. Another idea invokes tapping an aquifer, a deep layer of porous rock carrying liquid water and protected from evaporation by a permafrost cap. Deep burial would have compressed the water within the pore spaces of the aquifer, perhaps creating pressures high enough to have ruptured the overlying permafrost cap, allowing the aquifer to drain rapidly. The actual source of the water could, in this manner, be both melted permafrost and water from an underlying aquifer.

The second of the distinctive flow features marring the planet's surface are *runoff channels*, which are much more like the drainage systems of rivers on the Earth. They consist of sinuous valleys fed by many tributaries. On Earth, such branching networks are formed when rain falls in higher elevations and then runs downhill, as small trickles gradually coalesce into larger streams, in turn converging to form a mighty river.

These dry riverbeds occur only within the heavily cratered southern highlands of Mars, and their restriction to this ancient terrain

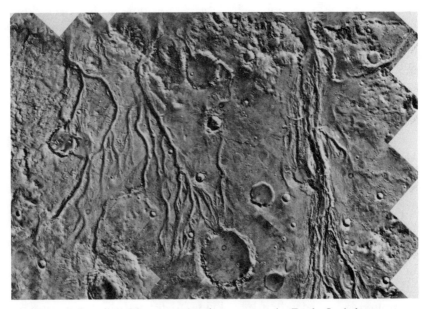

Branching valleys look like networks of streams on the Earth. Such features may have formed when rainwater coalesced into streams and then rivers in the distant martian past. (*Jet Propulsion Laboratory.*)

suggests that they must have been formed long ago. Indeed, the ages of these valleys, as estimated from abundant superimposed craters, are in the neighborhood of 3.5 to 4.1 billion years. If rainfall was necessary to make these sinuous channels, Mars back then would have been a very different place from what it is today. Billions of years ago, it may have been a cloudy and rainy world like the present-day Earth.

AN INVENTORY OF MARTIAN WATER

Just how much water did Mars have then, and what has happened to it? These questions are not easy to answer. Heinrich Wanke and Gerlind Driebus of the Max Planck Institute in Mainz, Germany, have used the compositions of martian meteorites to estimate that slightly over 2 percent of the mass of Mars, a fairly large proportion, was originally water. However, they believe that most of this water combined quickly with metallic iron during planet formation, forever locking the oxygen into rust and allowing the hydrogen to escape. The few water molecules that were left behind in the martian mantle after this reaction was completed amount to only a few millionths of its mass. Could this tiny amount of mantle water account for the observed features?

The Earth was not born with its swirling gaseous envelope and oceans of water. Like Mars, its water and other volatile constituents were trapped initially in rocks located within its interior. The gases were purged from the mantle by heating and dissolved in lavas, which carried them to the surface and exhaled them. This process is still going on today, as anyone who has stood near an active volcano and smelled its gaseous belching can attest. Over time, degassing of the Earth's interior produced a thick atmosphere that, when sufficiently saturated with water vapor, condensed to form the oceans. A similar process must have produced an atmosphere at one time on Mars, and possibly even shallow oceans. If all of the water estimated by Wanke and Driebus to be in the martian mantle were outgassed and condensed to liquid, it would make an ocean 130 meters deep covering the entire planet. This figure would be lower if, as seems likely, outgassing of the martian mantle was not complete.

Another way to determine the amount of water on Mars is by estimating the amounts of gases other than water vapor released from the martian interior. Noble gases are chemically inert, they do not freeze, and they are too heavy to escape Mars' gravity. Once released into the atmosphere, they are effectively there for the duration. We can determine with reasonable accuracy the ratio of the amounts of various noble gases to water on the Earth. If we assume that this ratio applies to Mars as well, we can then calculate the amount of water from a measurement of noble gases released into its atmosphere during its history. A failed Soviet Mars lander made the first attempt to analyze noble gases in the martian atmosphere. From telemetry received during its descent to the martian surface, scientists deduced that the atmosphere contained astoundingly large amounts of such gases, which translated to an enormous amount of water—the surface of the planet should be partly drowned under oceans, as it is on the Earth. The Viking landers, however, discovered that the previous measurements were wrong; instead, they measured an atmospheric noble gas content corresponding to a smaller amount of outgassed water, which, when condensed, would produce a layer on the surface on the order of 100 meters deep. Estimates based on measurements of the isotopic composition of hydrogen and nitrogen in the atmosphere are equivalent to global water depths varying from 3 to 120 meters. These results are in fairly good agreement with the estimate of Wanke and Dreibus, considering the uncertainties in all these calculations. By way of contrast, the Earth has outgassed enough water to cover its surface uniformly to a depth of 2,700 meters.

HIDE AND SEEK

A global ocean a few tens to 100 meters deep is no wading pool, and Lowell's martians or a visiting astronaut could get pretty tired treading water on such a planet. But of course that wouldn't be necessary, because the surface of Mars has been a parched desert for a billion years or more. Where could all this water have gone?

It is certainly not present now in the martian atmosphere. A typical atmospheric value for water vapor measured near the Viking lander sites is only about three hundredths of a percent by weight. The precise amount of water vapor varies with season and location, as small quantities move from pole to pole with the changing seasons, but even at its soggiest the atmosphere contains very little water. In fact, if all of the water vapor in the entire atmosphere were condensed, frozen, and packed together, it would comprise an ice cube measuring just 1.3 kilometers on a side. Spread uniformily over the globe, this would make a minuscule layer only a few billionths of a meter thick. Still, condensation of atmospheric water actually occurs on Mars at special times. At night the thin atmosphere becomes nearly saturated with water vapor as its temperature is lowered. On cold winter evenings, a very thin rime of frost can appear on the surface, only to evaporate in the morning Sun.

Likewise, gradual escape of Mars' tenuous atmosphere to space cannot explain the missing water, although some modest amount of water vapor has undoubtedly been lost by this mechanism. At an altitude of forty kilometers above the martian surface, water is broken down into smaller molecules and atoms by photochemical reactions. These new molecules and atoms are lighter than water and eventually escape from the planet's gravitational grasp. Even over the 4.5 billion years of martian geologic history, however, this loss has not been appreciable. So if the water is not now in the atmosphere and has not escaped from the atmosphere in the past, it must still be around. But where?

Some water must be stored in the ground, hidden within particles of martian soil. The rusty red color of the soil, and of the planet itself, suggests that the surface of Mars had undergone extensive chemical alteration. We do not know the mineralogy of the surface materials, but the two *Viking* landers have provided some clues. Each lander was equipped with a three-meter-long retractable arm for scooping up dirt and an instrument that analyzed its elemental composition by bombarding it with X rays. The chemistry of surface materials measured at both sites was identical, suggesting that well-mixed, wind-blown silt blankets the entire globe. Its composition is unlike any soil or rock on the Earth's surface, so we can only guess at the identities of the minerals that comprise it. The red coloration is undoubtedly due to oxidized iron, and magnets on the *Viking* sampling arms attracted small particles of iron oxide, probably the mineral maghemite. This oxide is apparently mixed with clays and other minerals. What is important in our search for the missing water is that some of these minerals contain water molecules bound into their crystal structures. In addition, whole water molecules may be physically attached to the surfaces of mineral grains in the soil.

A more important hiding place for water on Mars, already alluded to earlier, is permafrost. The occurrence of a widespread underground reservoir of ice on Mars is suggested by a number of peculiar landforms. One that I have already mentioned is chaotic terrain, formed when the icy basement undergirding the ground melts and rapidly drains away. Another unusual type of terrain seen in martian photographs has been described as fretted, implying that it has been gradually gnawed away. Like chaotic terrains, these fretted areas formed by removal of underlying permafrost supports, but in this case the mining and collapse were gradual rather than sudden. Like blocks of cheese that are progressively whittled away a sliver at a time, towering escarpments have slowly receded as ice exposed on the cliff walls evaporated.

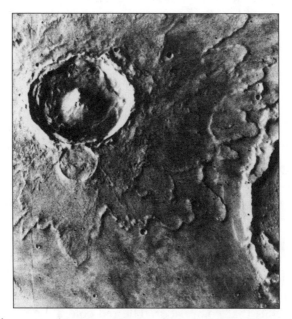

Ejecta blankets around impact craters on Mars look like splattered mud. The meteor that formed this crater probably melted underground ice, which mixed with soil to produce a surge of mud. This crater, named Yuty, is some twenty kilometers across and has a prominent central peak, possibly formed by the explosive expansion of ice as it was vaporized. (*Jet Propulsion Laboratory.*)

Perhaps the most intriguing indication of subsurface ice on Mars comes from its impact craters. Many craters there have a bizarre appearance, unlike the impact features on any other solar system body. Craters on the Moon and other planets are surrounded by material heaved outward in ballistic trajectories, like cannon shells. Near the crater rims, the ejecta blankets are coarse and blocky, and farther outward, the ejected rubble grades into finer materials that ultimately merge into dimpled fields of small, secondary craters and jagged rays. In contrast, the ejecta aprons around many martian craters look rather like splattered mud puddles. They consist of raised, rounded lobes, apparently formed from surges of wet mud. Geologists mapping Mars have applied the term rampart craters to these features, to signify that the excavated bowls sit atop pedestals of their own making. The splashed, muddy ramparts probably arise from subsurface ice that was melted by the force of the impacting meteor and mixed with soil. Rampart craters occur all over Mars, suggesting that permafrost may lie just under the surface almost everywhere.

ICE AT THE POLAR CAPS

But not all martian water plays hide-and-seek. At each of the poles, frozen water stands naked for all to see, just as it does on the Earth. In fact, the polar caps are likely to be a major repository for water on Mars. Polar regions contain three different geologic units atop one another. From bottom to top, these are the layered deposits, perennial ice, and seasonal frost caps.

The layered deposits consist of delicate, pastrylike beds of light and dark material, piled layer upon layer like a petrified baklava. The darker beds are probably windblown dust, and the brighter layers are thought to be water ice frozen out of the atmosphere. These deposits are cut by deep troughs that give the polar regions a distinctive swirling appearance. At each pole, the troughs spiral outward, counterclockwise in the north and clockwise in the south. The origin of these swirling valleys is not understood, but they provide a breathtaking view of the delicate laminations within the layered deposits. The top surfaces of the layered deposits, where exposed, are very sparsely cratered, so they are thought to be very young.

A layer of perennial ice overlies the layered deposits. As its name implies, this ice remains in place throughout the martian year. The diameter of the perennial cap in the north is 1,000 kilometers, almost reaching to the edge of the layered deposits, but it measures only 350 kilometers across at the southern pole. Its thickness at either pole has not been well determined, but it is probably on the order of a few meters. The compositions of the perennial ice caps have been a somewhat contentious subject. Prior to the *Viking* missions, planetary scientists debated whether they were composed of frozen water or frozen carbon dioxide (dry ice). This argument has been at least partly settled by temperature measurements over the perennial ice caps made by the *Viking* orbiters. At the north pole, temperatures are cold enough that water vapor in the thin martian atmosphere freezes out. This measurement, plus the observation of relatively large amounts of water vapor in the atmosphere over the north pole in the summer, indicate that the northern perennial ice cap must be frozen water. It is forty-five degrees colder at the south pole, which should allow freezing of carbon dioxide, but water might be an important constituent of the southern perennial ice cap as well. Nevertheless, because both perennial ice caps are relatively thin and together cover less than 1 percent of the martian surface, they are unlikely to be a major source for water on the planet.

A seasonal frost cap completely covers each polar region during its winter season. The rotational axis of Mars is inclined at about the same angle as that of the Earth, so both planets experience similar seasonal changes. In summer, the transient frosty lid shrinks as it evaporates, and clouds of gas migrate to the opposite pole where they recondense to form a new frosty covering. Temperatures during the martian winter at both poles are cold enough to freeze carbon dioxide, which is what the seasonal frost caps are made of.

The current seasonal variation in temperature represents just one of many periodic changes in the martian climate. Cyclical changes also affect the Earth. At some time you have probably seen artists' renditions of Pleistocene hominids, rugged fellows standing knee-deep in snow, bedecked in furs and brandishing clubs at woolly mastodons. These were residents of the last ice age, a time when glaciers advanced and our ancestors huddled against the cold in caves. The Earth has experienced four such chilly periods within just the last 2 million years. Their cause is uncertain, but most geologists believe that they resulted from minor, cyclic variations in the Earth's orbit. Mars also undergoes variations in its orbital motion that might likewise affect its climate over periods of 100,000 to 1 million years. The orbital oscillations that Mars experiences are similar, but more drastic, than those that have been suggested to have caused ice ages on the Earth. The details of such cosmic controls on temperature and thus the fate of water on Mars are not well understood, but they probably play a crucial role.

IMPLICATIONS FOR LIFE ON MARS

Mars is certainly not bone dry, but at the present time its water is mostly locked up in the form of ice at the poles and permafrost under the ground. Like the plight of the Ancient Mariner, there's water everywhere and not a drop to drink. Much of the interest in martian water comes from the fact that we, like Percival Lowell, cannot imagine any kind of life form existing without it. An average person living for seventy years will, in his lifetime, drink and recycle a million times his weight in water. And we humans are not atypical in terms of our water requirements, at least for the terrestrial life forms with which we are familiar. There are no watering holes on Mars, but there are probably reachable sources of ice to melt into life-sustaining fluid.

The *Viking* landers, designed primarily to search for life on Mars, were purposefully targeted to places where water might be obtained most easily. *Viking* 1 was originally slated to land in a region called Chryse Planitia, near the confluence of four channels scoured by running water in the distant past. Late in the mission this landing site was deemed too risky, however, because of the roughness of the surface. A more eastwardly location, still in Chryse but far removed from any visible evidence of water, was then selected. Once it had set down, *Viking*'s cameras panned across a dusty, reddish brown vista strewn with boulders and drifts of sand—and not a drop of moisture, nor any sign of life. Like its twin sister, *Viking* 2 had a last-minute change of landing site while still in orbit around Mars. Its original target was a more northerly location at which, it had been speculated, small quantities of liquid water might occur. Significant amounts of water vapor had been detected from orbit over this site, at least during the warm summer months. However, there was no way to verify how rough the landing surface might be at this location, so mission controllers substituted another site known to be smooth. *Viking* 2 set down in Utopia Planitia, at the same latitude as the prime site, about a month after its sister craft had landed. Despite its hopeful name, this site proved to be as dry and seemingly lifeless as *Viking* 1's resting place.

Although some scientists have gone to great pains to emphasize that the *Viking* biology experiments have not definitively ruled out the possibility of microscopic life forms on Mars, it now seems probable that animals and plants living on the Earth are very much alone in the solar system. Even if *Viking* had succeeded in finding life in the martian soil, I suspect that Percival Lowell would have been pretty disappointed with the outcome. He wished for an intelligent civilization, capable of constructing a globe-encircling system of canals to supply its drinking water and to grow its vegetables. The martian reality is very different. Nevertheless, we have discovered that abundant liquid water existed on Mars in the past, and frozen water is there now. Where there is water in some form, there is always hope for life. Perhaps future astronauts will find the fossil remains of extinct organisms that once thrived in a past wetter climate, or possibly primitive bacteria that now live underground and subsist on permafrost. And even if Mars proves to be sterile, it may someday provide a sanctuary for life forms transplanted from our own planet. Earthlings bent on colonizing a new world would certainly find it easier to terraform a portion of Mars than the Moon. After all, water, a necessity for the maintenance of life, is already there.

Some Suggestions for Further Reading

Arvidson, R. E., Binder, A. B., and Jones, K. L. 1978. "The Surface of Mars." *Scientific American*, vol. 238, no. 3, pp. 76–89. Describes features of the martian surface, as viewed by *Viking* spacecraft and various *Viking* experiments.

Batson, R. M., Bridges, P. M., and Inge, J. L. 1979. *Atlas of Mars*. Washington, D.C.: NASA Special Publication 438. Crisp photographs and maps showing water-sculpted terrains on Mars.

Carr, M. H. 1981. *The Surface of Mars*. New Haven, CT: Yale University Press. This is an excellent and authoritative summary of martian geology; it includes a detailed account of water and ice, and the landforms produced by running water.

Hamblin, W. K., and Christiansen, E. H. 1990. *Exploring the Planets.* New York: Macmillan Publishing Company. A well-written introductory text for planetary geology; Chapter 5 describes the geology of Mars and has an especially good description of the *Viking* landing sites.

Sagan, C. 1980. *Cosmos.* New York: Random House. Chapter 5 of this popular best-seller has an interesting account of the history of martian canal mania.

Squyres, S. W. 1984. "The History of Water on Mars." *Annual Review of Earth and Planetary Sciences*, vol. 12, pp. 83–106. A technical review of planetary research leading to an understanding of martian water and ice.

Keplerian Litter

―――

The Mineralogy of Asteroids and Ring Particles

The same laws of physics that control the celestial motions of gigantic planets would also rule the orbital behavior of cat litter. This is a significant point, even though there are no feline astronauts. Cat litter–size particles are abundant in the solar system, part of a continuum of small objects ranging from microscopic dust motes to boulders that could just squeeze through a basketball hoop or a garage door. Billions of such pebbles form rings about the giant planets, and myriads of larger but still modest-size bits and pieces constitute a girdle of asteroids around the Sun. The larger asteroids themselves bump and grind against each other, over time generating a cloud of pulverized dust that is visible from Earth as the zodiacal light, a faintly lit band seen just before sunrise or after sunset.

PARTICLE ORBITS

Our current understanding of particle motions in planetary rings and in the asteroid belt derives mainly from the pooled research of an unlikely pair of eccentrics, Tycho Brahe and Johannes Kepler. To say that Tycho was flamboyant would be an understatement. He was perhaps the most unconventional and controversial figure of his day, at least in his native Denmark. He even looked the part, sporting six-inch-long handlebar moustaches and, in place of his nose, which he lost in a duel, a homemade prosthesis of gold, silver, and wax. A nobleman by birth, he flouted the customs of his class, even to the point of taking a commoner as his common-law wife. His drinking parties were legendary, and he had rancorous disputes with nearly everyone who knew him. But there was another facet to this complex individual. When he was not shocking the sensibilities of Danish high society, Tycho busied himself making detailed observations of the motions of planets. And very good observations, I might add. This work eventually gained him an appointment as mathematician to Rudolf II, Emperor of the Holy Roman Empire and astrology buff. His new post also necessitated that Tycho move to Bohemia, much to the relief of his Danish kinfolk. He carried along voluminous records of his many years of astronomical observations, to facilitate his astrological predictions.

In 1600 this burly, overbearing, and abrasive dandy chose as his assistant a squirrelly, threadbare, and accident-prone Johannes Kepler, a fateful decision that was to revolutionize astronomy. At Tycho's untimely death (some say suicide) the next year, his astronomical data fell into the hands of the young assistant. Kepler, like his mentor, had a few loose screws. A devotee of religious mysticism, he was fascinated by the musical notes supposedly emitted by the planets as they moved harmoniously in their orbits. Fortunately, his other scientific contributions were more substantive. After decades of wrestling with Tycho's data, Kepler finally recognized that planets follow elliptical orbits, which requires that they move faster when they are nearer the Sun. Although we now describe such motions as Keplerian, this triumph also belonged in part to Tycho, whose careful observations formed the basis for Kepler's laws of motion.

Although Tycho and Kepler were concerned primarily with the orbits of planets, their insights also provide information about the motions of much smaller objects. A cloud of particles orbiting a central object, whether the Sun or a planet, eventually flattens into a thin disk, with those particles nearer the object traveling at higher velocities than those farther away. As the more rapidly moving inner particles overtake neighbors just slightly farther out, they may bump against them. This jostling tends to retard the inner particles and accelerate the outer particles, a phenomenon known as Keplerian shear. The slowed inner particles spiral farther inward to occupy orbits closer to the central body, and the accelerated outer particles move still farther away. In other words, the disk spreads laterally in both directions. The form of Saturn's glorious ring system, some 200,000 kilometers wide and barely a few tens of meters thick, dramatically illustrates the effect of Keplerian shear. The asteroid belt, a thin band of objects several hundred million kilometers wide, is a larger version of the same phenomenon.

The motions of small objects such as ring particles and asteroids are affected by other nearby objects as well. As the disk spreads, certain areas are swept clear of particles. To understand how this happens, imagine the disk as a racetrack. A fast car hugging the inside rail makes two laps in exactly the same time it takes a slow car on the outside of the track to make one lap. Now let's redefine the fast car as a ring particle and the slow car as a moon, both in orbit about a planet. The orbital periods of the two objects are whole-number multiples of each other, a condition known as resonance. Whenever the orbiting particle is adjacent to the moon, it is attracted more strongly by the moon's gravity. The frequent gravitational tugs are additive and may distort the particle's orbit to the point where its motion becomes chaotic, causing it to swerve off the racetrack. The Cassini division, a relatively empty zone in Saturn's ring system, demonstrates just this effect. Particles within this zone line up adjacent to the satellite Mimas on every second revolution, and over time this region of the ring has become largely depopulated by Mimas' gravitational pull. Analogous empty zones, called Kirkwood gaps, in the asteroid belt are caused by resonances of asteroids at these orbital locations with Jupiter.

Other, non-Keplerian forces also can modify the motions of particles, causing them to spiral continuously inward toward the bodies about which they orbit. For example, the innermost particles of Saturn's rings scrape the top of its atmosphere, causing them to slow and fall inward and thereby sweeping clean a region up to 7,000 kilometers altitude adjacent to the planet. Dust grains in the asteroid belt similarly work their way inward toward the Sun, because they lose energy through interaction with sunlight.

ORBITS AND MINERALOGY

The motions of planetary ring particles, asteroids, and asteroidal dust are immensely complex, and I have scarcely touched on this topic. Suffice it to say that there are many dynamic mechanisms that might drastically alter the Keplerian orbits of this cosmic litter over the course of solar system history. But have they really done so? Have the orbits of small objects been so thoroughly scrambled and the bodies so scattered about that they no longer reside near the places where they formed?

This may seem like a purely astronomical, rather than geologic, question, and you may be wondering why we are considering it in this book. A key to answering this question lies in the mineralogy of these small objects, and therein lies the connection. Geologists define *mineral* as a naturally occurring, inorganic substance with a precise crystal structure and a limited chemical composition. Most people recognize the exquisitely faceted crystals in museums and rock shops as minerals, but may not appreciate that the mundane gravel in their driveways is composed of the same substances. Atoms in the mineral grains of these rocks are organized into precise crystal structures, even though the interlocking grains do not have crystal faces. Minerals are all around us, in some cases even within us. Our teeth, for example, are crystals of calcium phosphate, the mineral apatite. But not everything is a mineral. A drop of water has no rigid crystal structure, a maple leaf is organic, and steel is man-made. All three fail to meet some aspect of the definition of a mineral.

Identifying the mineralogy of planetary ring particles and asteroids can help us decide whether they have been flung wildly about, because the minerals that comprise them should differ depending on where they originally formed. If the orbits of cosmic litter have not been redistributed, we might expect that the mineralogy of ring particles and asteroids should vary with solar distance, in the same manner as do the compositions of planets. The inner planets are composed mostly of silicate and oxide minerals and metal alloys, in most cases similar to those comprising the planet on which we live. In contrast, the constituents of the outer planets, mostly liquefied volatile compounds such as molecular and metallic hydrogen, are not really minerals at all, at least as we have defined them. However, the constituents of the giant planets are stable as liquids only because they are under enormous pressures in the interiors of these massive bodies. In objects as small as asteroids or ring particles, these same constituents would form ices. Naturally occurring ices are crystals with limited compositional ranges and thus do meet the definition of minerals. So it may be convenient for us to visualize the mineralogy of the giant planets as mostly ices, although that is not strictly correct. These planets also have rocky cores, so silicate minerals also may play a role.

THE ASTEROID BELT

Let us first focus on asteroids, to see if they follow the same mineralogic progression (silicates, oxides, and metals to ices) as planets as we go outward from the Sun. Determining the mineralogy of asteroids is a daunting task, but a few maverick astronomers have risen to the challenge. Viewed through a telescope, an asteroid is a mere pinprick of light, distinguishable from distant stars only by its movement. The very name asteroid means "starlike." However, unlike stars, asteroids do not shine by their own light, but by reflected sunlight. One basic observation that can be made is how well or poorly an asteroid reflects light, a property called albedo. The brightest asteroids reflect nearly half of the incident sunlight, whereas the darkest asteroids reflect almost none. Albedo is affected by how well ground up are mineral grains on asteroidal surfaces, but it also depends on the identity of the minerals and the presence of other, nonmineral constituents such as organic compounds.

Certain minerals also may produce subtle changes in the reflected light, a familiar effect we recognize as color. As a prism demonstrates, a sunbeam can be split into a spectrum of light vibrating at different wavelengths. When sunlight impinges on an asteroid, some wavelengths of light are preferentially absorbed and others reflected. If, for example, the blue (short-wavelength) end of the spectrum is absorbed, we see only the reflected complementary colors at the red end of the spectrum. So, when astronomers refer to asteroids as being red or blue, they do not mean that these objects actually have the color of an apple or the sky. They are saying that the light reflected from certain asteroids contains a higher proportion of short (bluish) or long (reddish) wavelengths than the incident sunlight. Using a combination of color and albedo, asteroid watchers have now classified asteroids into more than a dozen distinct groups.

We can begin to decipher the mineralogic meaning of these groups by squinting at the details of each asteroid's mirrored reflection. The measured spectra of asteroids are commonly plotted so that we can visualize how well they reflect or absorb different wavelengths of light. Some asteroids, such as shiny Nysa, have high albedos, whereas others, such as coal-black Ceres, do not reflect much light. The spectra of many bodies, such as Vesta and Eros, show irregular dips and peaks that represent the absorption or transmission of particular wavelengths of light. Some, such as Ceres, absorb most of the light at short wavelengths and thus are reddened. Still others, such as Nysa, have spectra that are relatively flat and featureless.

The astronomers who measure asteroid spectra are, in a very real

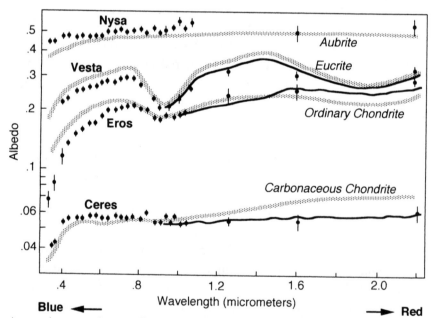

Asteroids, arrayed vertically in this diagram according to their albedos, reflect or absorb different wavelengths of light. The spectra of sunlight reflected from asteroids are similar to the spectra of crushed meteorite samples, suggesting that they consist of similar minerals. The asteroid measurements are shown by continuous black lines or points with error bars, and meteorite measurements are indicated by gray lines.

sense, mineralogists. The shapes of these spectra are controlled by the interaction of light with the atoms in crystals, and in some cases they provide diagnostic clues as to the mineralogic identity of those crystals. For example, a prominent valley in Vesta's spectra is due to absorption of a certain wavelength of light by iron atoms in the mineral pyroxene, a magnesium-iron silicate. Still other spectral features result from absorption of light by water molecules bound within crystals. Ceres absorbs infrared light, which has just the right energy to send water molecules into a Jazzercise frenzy of stretching and bending. From this absorption, asteroid astronomers have deduced that Ceres is composed partly of clays, silicate minerals that contain water bound into their structures.

We can check inferences about the mineralogy of these faraway objects in laboratories on the Earth by measuring the spectra of sunlight reflected from powdered rock samples with known mineralogy. Meteorite spectra provide particularly interesting comparisons, because most meteorites were derived from asteroids. Similarities, both in terms of albedo and shape of the spectral curves, suggest that certain meteorites and asteroids are composed of the same minerals. Bright asteroids apparently consist of iron metal and the silicate minerals olivine and pyroxene. Asteroids with modest albedos are similar to carbonaceous chondrites, which contain water-bearing clays, magnetite, and organic compounds. No known meteorite analogs exist for the darkest asteroids, but their spectra suggest abundant ice and organic compounds, along with dry silicate minerals. Spectral comparisons are even used to argue that various types of meteorites were derived from a specific asteroid or group of asteroids. However, asteroid spectra are more variable than the spectra of known meteorites. Apparently, the Earth has been pelted over the last few centuries by only a limited and highly biased sample of asteroidal materials.

Over forty years elapsed between the discovery in 1920 that asteroids differ in color and albedo, and the first serious attempt to use these properties to interpret their mineralogy. It took another thirteen years before it became apparent that asteroid mineralogy varies in a systematic way with solar distance. Journalist Edward R. Murrow might just as well have been describing the pace of this scientific discovery when he quipped, "The obscure we see eventually. The completely apparent takes a little longer."

To see how asteroid spectra change with orbital location, let's assume that we are aboard a spacecraft traveling outbound from the Sun and that we take snapshots of the asteroid population as we pass by a few locations. At a distance of 2 astronomical units (A.U.), just a little past Mars, we find that virtually all of the asteroids we encounter are members of groups with high albedos. Some of these, such as Nysa, have flat spectra, and others, such as Vesta, have curvaceous shapes. At a distance of 2.3 A.U., the number of bright asteroids begins to decrease, and asteroids with modest albedos and infrared absorption bands increase until peaking in abundance at 3 A.U. Moving farther outward, the moderately bright asteroids are gradually supplanted by dark, reddened asteroids. The actual pattern is somewhat fuzzier and more complex than this simple picture, but, in general, each asteroid spectral group has its own special preserve within the asteroidal zoo.

Even after the special orbital locations of different asteroid groups were recognized, it took some time to recognize the mineralogic significance of this pattern. The asteroid groups that dominate the inner part of the asteroid belt apparently are composed of minerals that crystallized from molten magma. Vesta, for example, is probably covered by hardened basaltic lava. Some other bodies in the inner part of the asteroid belt have spectra similar to iron meteorites. These asteroids are probably remnants of larger bodies once covered by lavas, but their rocky mantles have been stripped off by impacts to expose metallic cores. Cores within asteroids, like the cores within planets, were presumably formed by wholesale melting and sinking of dense, molten metal.

In contrast to these once-melted objects, asteroids with modest albedos farther out in the main belt have fundamentally different mineralogy. These bodies apparently contain clays and oxide minerals formed through alteration by liquid water. These asteroids once might have been mixtures of ice and dry rock that were then mildly heated so that the ice melted and reacted with the silicate minerals. Still farther away in the outskirts of the main belt are the darkest objects, thought to contain ice mixed with dry silicate minerals and organic compounds. These "iceteroids," in which the ice apparently never melted, are somewhat similar to comets but have less ice.

One inference we can draw from this asteroidal geography lesson is that a "snow line" separates rocky or metallic asteroids from bodies that now consist of, or formerly contained, both ice and rock. The occurrence of ice in the asteroid belt is obviously related to distance from the Sun, just as it is in the planets. Overlaid on this pattern of varying amounts of rock and ice are the effects of heating, which apparently diminished in intensity with solar distance. The rocky bodies nearest the Sun were partly or completely melted, resulting in the formation of metallic cores in their interiors and the eruption of silicate lavas on their surfaces. The locale for *The Little Prince*, a fanciful story of a boy who inhabited an asteroid not much bigger than himself, but with three knee-high volcanoes, would have been one of these melted objects. Temperatures farther out past the snow line were lower, inadequate to melt rock but sufficient to melt ice. The resulting liquid water reacted with dry rock to form clays and oxide minerals. At the outer reaches of the asteroid belt, temperatures were too low even to melt ice, so iceteroids there have survived intact.

We do not know what caused asteroid heating, but the relationship between heating intensity and distance from the Sun suggests some solar involvement. One possibility is heating by electrical induction, a process suggested to have been caused by a rapidly rotating Sun expelling large quantities of matter in a ferocious solar wind. Astronomers have observed young stars doing just that at an early stage of their development, and it seems plausible that our Sun went through a similar evolution. In induction heating, the magnetic field of the solar wind was converted into electrical currents inside planets and asteroids in its path, and resistance to the flow of these currents generated heat. The intensity of the solar wind would have declined with distance from the Sun, and so consequently would the amount of heating. However, induction heating is a controversial idea, and there are other ways of heating asteroids that could explain the variation in intensity with orbital distance as well. For example, heating may have been caused by the rapid decay of short-lived, radioactive isotopes. The observed heating pattern in the asteroid belt might then reflect the sequential accumulation of matter into asteroidal bodies, first nearer the Sun and then farther away. Those asteroids that accumulated earlier received a higher proportion of "live" isotopes and so were heated to higher temperatures than the asteroids that formed after much of the radioactive decay had already occurred.

So, the asteroid belt was originally stratified into a dry, rocky inner zone that was strongly heated, a soggy middle zone that was mildly heated, and an icy outer zone that was virtually unheated. This original zonal structure of the asteroid belt is still preserved today. In other words, the orbits of most asteroids have apparently *not* been scrambled. A little orbital reshuffling has occurred, to be sure, as evidenced by the removal of asteroids from Kirkwood gaps and the periodic collision of the Earth with asteroid fragments in the form of meteorites. However, many of the asteroids left behind in the main belt appear to reside in their original cosmic neighborhoods.

PLANETARY RINGS

Does this conclusion also hold true for litter in other regions outside the asteroid belt? The mineralogy of planetary ring particles offers another opportunity to test whether the orbits of small objects have been scrambled.

How planetary ring systems form is not exactly clear. We do know that all rings lie within the *Roche limits* of the planets they encircle, the regions close to these planets where tidal forces that tear bodies apart exceed the gravitational force holding them together. Any dispersed matter within the Roche limit would form a ring system rather than clump together to form satellites. These planetary hula hoops could have formed at the same time as their host planets, or later, when moons or wandering asteroids strayed too close and were wrenched apart by tidal forces. The latter situation is an example of orbital scrambling, and we would not expect the mineralogy of rings formed late in solar system history to follow the same sequence as the mineralogy of planets and asteroids.

Saturn's ostentatious rings make it the most readily recognized object in the solar system. As viewed close-up by *Pioneer* and *Voyager* spacecraft, this ring system is breathtaking. It is so thin that its appearance alters drastically with different viewing angles. The concentric rings have been labeled with letters (A through F) in order of their discovery, so their designations indicate nothing about their positions. The innermost D ring is normally invisible, appearing only in long-exposure photographs from nearby spacecraft. The tenuous

The extraordinarily complex rings of Saturn, as revealed by *Voyager* spacecraft. The rings have different appearances when viewed from different angles. (*Jet Propulsion Laboratory.*)

ringlets that form this band spiral into the planet because of atmospheric drag. Next outward is the C ring, marked by regularly spaced light and dark areas. The B ring is the widest, thickest, and brightest of Saturn's halos. In addition to its thousands of ringlets, this region is characterized by bizarre spokes. The Cassini division separates the B and A rings, though *Voyager* images indicate that this boundary is far from the void that was originally described. Compared to others, the A ring is relatively featureless and transparent. Outside of that lies the F ring, a narrow band of material with unruly kinks, braids, and knots. A faint E ring, visible only when viewed edge on, extends from the F ring exterior some eight Saturn radii outward.

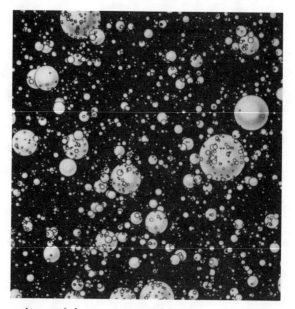

This artist's rendition of the appearance of Saturn's A ring close up shows that it consists of ice chunks varying in size from marbles to beachballs. *Voyager* measurements were used to estimate what a flight through this ring might look like. *(Jet Propulsion Laboratory.)*

The sharply defined rings and ringlets around Saturn require some special explanation, since they imply that some force is resisting the effect of shear that causes the rings to spread. The sharp rings can best be explained by the presence of shepherding moonlets, which act to focus the orbiting particles into tight bands through resonance. But many other features of ring systems, such as the occurrence of ringlets within the Cassini division or braids within the F ring, defy explanation, at least at our current level of understanding.

The sizes of Saturn's ring particles vary from microscopic specks to automobile-size boulders. They are quite bright, and their high albedo explains why we are able to see Saturn's rings from Earth. The billions of particles are composed of water ice, a blizzard organized into militarily precise formations marching along at ten kilometers per second. The same spectral technique that identified water molecules on asteroids has been applied to ring particles, demonstrating the presence of frozen water. However, these particles are apparently not ice alone. Pure ice is white, but the rings are various pastel shades of tan, brown, and even red, suggesting that the particles are colored by impurities. The individual rings have distinctly different colors and sizes of particles, so that compositional distinctions have been maintained in different orbits, in much the same way that they have within the asteroid belt.

We have known of the existence of Saturn's rings for more than four centuries, but the ring systems around the other giant planets are recent discoveries. Jupiter's rings were discovered in 1979 by the *Voyager 1* spacecraft, which was programmed to take photographs along the planet's equatorial plane while traversing through it. To almost everyone's surprise, *Voyager* detected a gossamer disk encircling Jupiter. Jupiter's ring system is invisible from Earth because it is composed of very tiny, dark particles. These particles are either dark stones or ice coated with black dust or organic matter.

The tiny particles cannot remain in orbit around Jupiter indefinitely. Their small sizes insure that eventually they will either be pulled into the planet's atmosphere or ejected out of the ring and spiral into the Sun. The particles may result from impacts into satellites that dislodge bits of matter and scatter them into the ring. The fact that Jupiter's ring system must be replenished constantly indicates that it is of recent origin.

The ring system encircling Uranus was an accidental discovery in 1977. Astronomers positioned themselves in the southern hemisphere in March of that year, to record the passage of Uranus in front of a star. Their intent was to determine precisely the diameter of the planet by measuring the length of time its passage obscured the star behind it, a phenomenon known as occultation. A team of American astronomers was poised in Australia, while their French counterparts watched from South Africa. Still others found a room with a view in a C-141 Starlifter aircraft, fitted with a telescope and flying holding patterns over the Indian Ocean. The airborne group focused its telescope, equipped with a sensitive light-measuring device, on the star and waited eagerly for the light signal to diminish as it passed behind the planet. But half an hour before the occultation was due to begin, the star unexpectedly began blinking, five times in all. The observers reasoned that small objects near Uranus were blocking the starlight and hastily notified their compatriots on the ground to look for similar effects. After the planet occultation, which lasted for twenty-five minutes, the various teams observed five more brief blinks, mirror images of those noted earlier. The only logical conclusion was that Uranus had rings! This discovery was independently confirmed the next year by other astronomers, who expanded the number of rings to nine. *Voyager* 2 subsequently imaged the rings when it reached Uranus in 1986, discovering more dainty strands and revealing many new details about this unusual system.

Uranus' rings are narrow and crisp, ranging in width from as thick as forty kilometers to as thin as one kilometer. Their shapes probably result from shepherding satellites, although only two have been discovered so far. The visible rings are composed of meter-size boulders to house-size bodies. They are as dark as soot, reflecting only a tiny fraction of the incident light. When viewed with the Sun at its back (as could only be done by *Voyager*), the ring system is revealed to contain numerous thin bands of fine, dark dust interleaved among the visible strands of boulders. The exact mineralogy of the dark matter that comprises the Uranian ring system is not known, but scientists believe these boulders and dust particles are rich in carbon, perhaps some sort of cosmic tar formed through the irradiation of methane ice. They also could be rocky material similar in composition to chondritic meteorites.

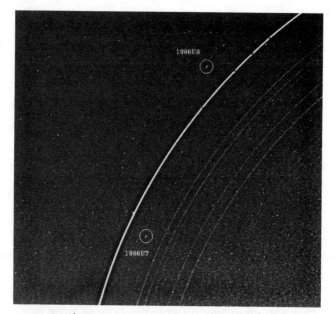

Particles comprising the most prominent ring of Uranus are kept in place by two tiny shepherding satellites, now named Cordelia and Ophelia. (*Jet Propulsion Laboratory.*)

Naturally, after ring systems had been found around all the other giant planets, many people supposed that Neptune would likewise have rings. However, attempts to find them by stellar occultations in the early 1980s were disappointing. A few patient observers finally struck paydirt in 1984, but their results were confusing. Occultation by Neptune's rings seemed to be a hit-or-miss proposition, leading these workers to suggest that the rings were incomplete arcs of particles. Finally, in 1989 *Voyager* 2 arrived at Neptune and provided the first close-up views of its rings. It turns out that three thin rings do completely encircle the planet, but the outer one is clumpy. Matter tends to be concentrated at certain locations along this ring, causing it to appear as a series of arcs. The origin of this strange phenomenon is unknown. These arc clumps may be recently fragmented satellites whose particles have not had time to form a continuous, smooth ring, or they may be particles shepherded into their current locations by unseen moons. Just as in the case of Uranus, Neptune's rings are composed of charcoal-black particles. The ring systems around both

The outermost ring around Neptune is discontinuous, consisting of clumps of orbiting ring particles that form arcs. (*Jet Propulsion Laboratory.*)

planets probably have similar compositions, but there is little information now available that might further constrain their mineralogies.

What do rings around the giant planets tell us about the orbital scattering of small particles? Certainly, forces acting on them have locally redistributed these objects. In fact, the ring systems of Jupiter, Uranus, and Neptune may be short-lived features, the result of a few stray moons wandering inside the Roche limits of these planets or of impacts that dislodged material from other, nearby satellites. Saturn's rings, though, may be something else entirely. Its ringed halo may have persisted for billions of years, perhaps even from the time of the planet's formation. The fact that the bright particles that comprise this ring system are water ice distinguishes them from the dark particles of other ring systems, demonstrating that orbital scattering has not transported this litter from planet to planet. The different colors of the various concentric rings may even signal that the mixing and scattering processes that have affected the tenuous ring systems of the other giant planets may not operate very effectively within the confines of this ring system.

A more fundamental conclusion is that all of the outer planet ring systems consist of materials that apparently formed in the outer solar system. The icy mineralogy of Saturn's hoops is certainly appropriate for its location out past the snow line, and the dark ring particles of the other giant planets appear to be similar to dark satellites around some of these bodies and dark asteroids in the outer part of the main belt. In a broad sense, these ring particles may be telling us the same thing that asteroids do—the gross compositional gradient present in the early solar system still persists today.

I still find it astounding that we can identify minerals on the surface of a small asteroid many millions of miles away and at least make some guesses about the mineralogy of tiny planetary ring particles. My geology students have a tough enough time identifying the large, distinctive specimens of common minerals they hold in their hands. Equally amazing is the idea that the mineralogy of these faraway particles, some the size of a grain of cat litter, bears a relationship to their orbital position. I don't know if Johannes Kepler was a cat lover, but I think he would be pleased to know that overturning a litter box in interplanetary space wouldn't necessarily make a mess of his orbital calculations.

Some Suggestions for Further Reading

Araki, S. 1991. "Dynamics of Planetary Rings." *The American Scientist*, vol. 79, p. 44–59. An authoritative paper on planetary rings, written for a general audience.

Beatty, J. K., and Chaikin, A., eds. 1990. *The New Solar System*, 3rd ed. Cambridge, MA: Sky Publishing Corporation. This excellent book contains pertinent chapters on planetary rings and asteroids. Chapter 12 by J. A. Burns tells all about ring systems, and Chapters 18 by C. R. Chapman and 20 by W. K. Hartmann give overviews of asteroids. All of these are particularly well written and illustrated.

Brahic, A., and Hubbard, W. B. 1989. "The Baffling Ring Arcs of Neptune." *Sky & Telescope*, vol. 77, pp. 606–609. A clear description of a complex feature in planetary rings.

Chapman, C. R. 1975. "The Nature of Asteroids." *Scientific American*, vol. 232, no. 1, pp. 24–33. A somewhat dated but very informative article on asteroid spectra, classification, and mineralogy.

Cuzzi, J. N., and Esposito, L. W. 1987. "The Rings of Uranus." *Scientific American*, vol. 257, pp. 52–66. A general-audience article on the discovery and nature of the ring system of Uranus.

Esposito, L. W. 1987. "The Changing Shape of Planetary Rings." *Astronomy*, vol. 15, pp. 6–17. An excellent review of planetary ring systems.

Lang, K. R., and Whitney, C. A. 1991. *Wanderers in Space, Exploration and Discovery in the Solar System*. Cambridge, England: Cambridge University Press. An excellent text, in which Chapter 7 gives information on asteroids and various chapters on specific planets provide more insights into ring systems.

From Sea to Shining Sea

Magma Oceans on the Moon and Planets

My daughter would add a few years to her age if she could, and my wife would subtract a few. Neither of their wishes is at all unusual; the young always seem to be in a hurry to grow up, and grown-ups lament their lost youth. I too could make good use of a younger body, fast enough to reach those pesky shots into the far corner of the tennis court and trim enough to fit into my old clothes. But, unlike most grown-ups, I might just be willing to add on a few extra years, with the proviso that they had been spent studying the first returned lunar samples. I have always envied those scientists, some of whom are only a few years older than I, who had the opportunity to participate in the *Apollo* sample program. The very first geologists to examine rocks from the Moon had an experience the likes of which the rest of us who practice this discipline will never know.

THE FIRST RETURNED LUNAR SAMPLES

The stories told by those privileged few still radiate a sense of wonder, and more than a few participants have become scientific folk heroes. Some of my most interesting lunch hours were spent huddled around a lab table at the Harvard-Smithsonian Center for Astrophysics, listening to my graduate advisor, John Wood, tell about the early *Apollo* program. Just after the *Apollo 11* capsule landed on Earth in 1969, Wood trekked to the Lunar Curatorial Facility in Houston to collect his allocated lunar sample in person. NASA had decreed that shipping pieces of the Moon via the postal service was out of the question. Wood took NASA's admonishment that the sample should not leave his person very seriously, to the point of actually borrowing needle and thread to sew the small sample vial into the pocket of his sport coat. On the way back to Boston, the commercial aircraft carried many such smugglers, and they joked among themselves that, should they crash enroute, newspaper accounts would be more interested in how many grams of lunar sample were destroyed than lives lost.

This time was a period of great student unrest, and Harvard Square was a place to be avoided. Wood recalls, with a laugh now, his concern that rampaging troublemakers might break into his laboratory to steal the lunar sample, this shining trophy of the government's space program. He and his colleagues even went so far as to label a fake lunar sample and place it in their safe.

In any case, Wood's reward for all these security precautions was one of mankind's first microscopic glimpses at a pinch of lunar soil. The novelty and euphoric thrill of handling a sample of the Moon, however, soon gave way to a crushing work schedule imposed by the short deadline for reporting research results. All this frenzied activity climaxed several months later at the inaugural Lunar Science Conference, with its palpable sense of excitement engendered by new revelations and by a newly heightened level of scientific competition. And if that were not enough, the conference participants were recast as interpreters, translating their normally arcane scientific prose into newsworthy tidbits for a buzzing swarm of reporters and television crews. This was heady stuff for those accustomed to the quiet isolation of the laboratory and the relative anonymity of the scientific researcher.

PROPOSAL FOR A LUNAR MAGMA OCEAN

In the tiny vial of lunar soil, Wood and his small band of collaborators saw something that almost everyone else missed. From the study of these minute sand particles, they conjured up a vast lunar ocean, inundating the entire Moon early in its history. This ocean, however, was nothing like the wet, blue seas we Earthlings know. It was a cauldron of roiling molten rock, a deep, globe-encircling lake of magma.

What could these geologists have seen in a few fragments of lunar soil that led to such a startling conclusion? Quite simply, they saw grains of feldspar (calcium aluminum silicate), also a common mineral on the Earth. What made the lunar feldspars so interesting was that they were usually welded to other feldspar grains, as if they were fragments derived from the breakup of rocks devoid of other minerals. A rock composed entirely of feldspar is called anorthosite.

These feldspars were discovered in soil collected from Mare Tranquilitatis, a large basaltic lava lake targeted as the *Apollo 11* landing site because its relatively smooth and featureless surface made it safe. Most of the dirt from this location was clearly derived from impacts into the underlying mare basalt. The large, blocky feldspar particles in the soil, however, were slightly different in composition and appearance from the tiny, needlelike feldspars in basalt. In Wood's view, these grains were interlopers, particles lofted from the distant lunar highlands by giant impacts. He surmised, then, that the highlands were composed primarily of anorthosite rock.

Lunar anorthosites are clearly igneous rocks, but magmas composed only of melted feldspars are improbable, and more realistic magma compositions do not crystallize feldspar alone. Wood and his collaborators proposed that the feldspars originally crystallized, along with other minerals, within a cauldron of basaltic magma but were concentrated by floating to the top. This buoyant scum eventually cooled and hardened into a "cumulate" anorthosite crust. At the same time, denser crystals of olivine and pyroxene (both iron-magnesium silicates) that formed along with the feldspars sank. The idea of crystals sinking and floating was not so controversial, as similar accumulations of crystals already had been documented in terrestrial magmas. What was mind-boggling was the scale on which the lunar process supposedly happened. Wood and his coworkers argued that the highlands everywhere, virtually the entire ancient lunar crust, were made of feldspar. Furthermore, they estimated that the buoyant feldspathic crust of the Moon was some twenty-five kilometers thick. The accumulation of such a feldspar layer required a lava lake of much greater

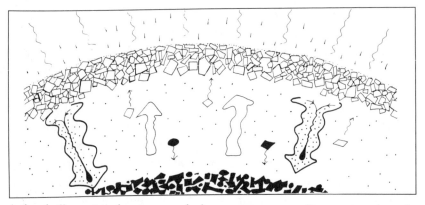

A sketch illustrating the concept of a lunar magma ocean. Dense minerals, such as olivine and pyroxene, sink to the bottom of the ocean, and lighter minerals such as plagioclase float. This model has been used to explain the thick, feldspar-rich crust of the Moon. *From D. Walker, Journal of Geophysical Research* 88 (1983): B17, *copyright American Geophysical Union, with permission.*

thickness, hundreds of kilometers, a magma truly of oceanic dimensions. They also suggested that the dark maria might be chilled samples of the magma ocean itself, exposed through holes in the feldspathic crust punched by impacting meteors.

Wood's idea was well received, but it needed the validation of actual highlands samples. It would be well over a year before an *Apollo* mission actually landed in the rough lunar highlands and recovered intact anorthosite rocks. The first such sample was collected at the foot of the Apennine Mountains by *Apollo 15* astronauts. During their second traverse on the lunar surface, David Scott and James Irwin noticed a large white rock perched conspicuously on a pedestal of darker dirt, almost as if to call attention to itself. The astronauts had been coached to look for rocks composed mostly of feldspar as potential samples of the ancient highlands crust, and they were ecstatic to find one so easily. This particular rock consisted of about 99 percent feldspar. On the assumption that this sample, later prosaically cataloged as "rock 15415," would provide long-sought clues about the Moon's origin, the astronauts enthusiastically dubbed it the Genesis rock.

Samples like this anorthosite shut the door on some alternative ideas about the Moon's ancient crust. Two decades prior to the Apollo program, the orthodox view was that the lunar highlands, so different

This large white rock is the Genesis rock, a sample of anorthosite collected by *Apollo 15* astronauts. It is presumably an intact piece of the ancient lunar high-lands crust. A thin slice of the Genesis rock, as viewed through a petrographic microscope, shows that it contains almost nothing but feldspar grains. The light and dark bands are the result of twinning of feldspar crystals. (*NASA Johnson Space Center.*)

in appearance from the Earth, must be a kind of meteoritic sediment accumulated onto a cold and rigid substrate. This idea was formulated on the popular belief that the heavily cratered highlands preserved an unblemished record of the Moon's earliest history. Later, during the time of the *Ranger* missions in the early 1960s, a belief sprung up that the lunar highlands had granitic compositions, like continents on the Earth. The *Apollo 11* samples upended both models, creating an empty stage for Wood's proposal of a magma ocean. This new idea, first published in 1970, was, despite its radical nature, readily embraced by most of the scientific community. It offered a unifying conceptual framework for understanding many different observations about the mineralogy, chemistry, and ages of the Moon's rocks.

That is not to say, however, that the hypothesis proposed by Wood and his colleagues was accepted without modification. Scientists love to pick away at ideas, gradually refining them until they meet all conceivable tests. Data on highlands samples such as the Genesis rock provided just such tests for the magma ocean hypothesis and resulted in a few changes. In the early 1970s, the required depth of the magma ocean began to shrink, as newer models allowed for more efficient cooling. The realization that impacts would have breached the tenuous crust periodically and caused it to founder, and the possibility of floating "rockbergs" of anorthosite, added to the complexity of a Moon drowned in magma. These were mostly variations on the original theme, but one aspect of the original hypothesis turned out to be manifestly wrong. The earlier suggestion that the lunar maria were outcroppings of congealed magma ocean exposed through crater windows clearly did not hold water. Radiometric age dates on mare basalts turned out to be as much as a billion years younger than the highlands, and so they could not have formed by quenching of this gigantic lava sea. Instead, these lavas must have formed by later remelting of the lunar interior or, more accurately, of the cumulate rocks at the bottom of the solidified magma ocean.

CHEMISTRY OF THE MAGMA OCEAN

One of the earliest arguments supporting the magma ocean hypothesis was that it readily explained a peculiar observation about the chemistry of highlands anorthosite rocks and of the basaltic lavas that later filled the maria. From a chemist's perspective, these two kinds of rocks are mirror images of each other. That is not hard to understand, if you recall that while feldspars floated in the magma ocean, denser minerals such as olivine and pyroxene sank to form an underlying mantle of different composition. Any element that was concentrated in the upper feldspar-enriched crust must therefore have been correspondingly less abundant in the feldspar-depleted mantle. This chemical difference is a hereditary trait, like blue eyes and freckles, capable of being transmitted to the next magmatic generation. Many millions of years later, the olivine- and pyroxene-rich lunar mantle was remelted to produce basaltic magmas that rose and ponded in large craters, now recognized as maria. Whatever distinctive chemical characteristics had been originally imprinted on the mantle were passed on to its basaltic offspring. So analyzing mare basalts provided a way to sample, at least indirectly, the deeply buried olivine- and pyroxene-rich rocks that formed at the same time as the upper feldspathic crust.

One of the early champions of the magma ocean hypothesis was Stuart Ross Taylor, a geochemist at the Australian National University.[1] Taylor's argument for the existence of a lunar magma ocean focused especially on the *rare earth elements*—lanthanum, samarium, dysprosium, terbium, gadolinium, and a handful of others. The names of these elements are not exactly household words, even to most geologists, but their abundances in rocks are very informative. The rare earth elements exhibit similar chemical behavior, and, consequently, they cannot be separated from one another easily during melting and crystallization. These elements do not fit comfortably into the crystal structures of most minerals crystallizing from a magma, because they have obtrusively large sizes and high electrical charges. Their presence in most minerals is tolerated only in trace amounts, with a singular exception: The rare earth element europium substitutes readily for calcium in feldspar crystals. A feldspar-rich cumulate crust should thus contain the lion's share of the magma ocean's europium, relative to other rare earth elements.

[1] Taylor is but one of three prominent Moon researchers who bear this surname, a triumvirate affectionately known within the lunar science community as the Father, Son, and Holy Ghost. This particular Taylor is the Holy Ghost, an exalted status conferred because of his seniority and notoriety as an author of several well-known books about the Moon.

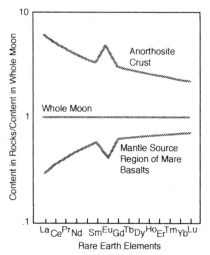

Samples of the highlands lunar crust, rich in feldspar, contain a larger quantity of the element europium (symbol Eu) than the other rare earth elements (whose symbols are shown at the bottom of the figure, in order of increasing atomic number). The lunar mantle, formed from minerals that sank in the magma ocean, was correspondingly depleted in this element. Rare earth element patterns for crust and mantle, compared to the whole Moon, are illustrated by curves. When the mantle was later melted to produce mare basalts, these liquids inherited the lower europium concentration of their parent rocks.

If feldspar concentrated europium, then the other minerals that crystallized from this same magma ocean, such as olivine and pyroxene, must have been depleted in europium. So, the top and bottom of a solidified magma ocean should have complementary abundances of europium and the other rare earth elements. The rare earth element concentrations in mare basalt, inherited from the accumulated olivine and pyroxene in the lunar mantle, show just this predicted mirror image. This complementary rare earth pattern between highlands and mare rocks is generally accepted as strong evidence supporting the former existence of a lunar magma ocean.

AN ALTERNATIVE MODEL

Throughout the 1970s the magma ocean model was the mainstay of lunar geology, but in the ensuing decade some former supporters began to waiver. As the *Apollo* booty of highlands rocks was examined in ever greater detail, it became clear that the ancient lunar crust was not just anorthosite. Another group of rocks, no more than half feldspar but with abundant pyroxene and olivine, was discovered in the highlands collection. Advocates of a magma ocean explained these rocks as products of later, unrelated pulses of magma that had intruded into the earlier, feldspathic crust. Other geologists, however, called for *serial magmatism*, by which they meant that the lunar crust had been built piecemeal as successive batches of magma ascended, solidified, and added their mass to the crust. Serial magmatism still required an impressive volume of magma, but not all of it was present at once—only a modest amount of melt had to occur within the crust at any one time. There was no need to appeal to a vast, sloshing magma ocean in this new model.

Serial magma pulses, like the magma ocean itself, would have had basaltic overall compositions, which means that when crystallized they would consist of approximately half feldspar and half olivine and pyroxene. Geologists call such rocks gabbros. In the case of a magma ocean, the olivine and pyroxene sank, leaving a crust of mostly feldspar. Like the magma ocean, crystallizing feldspars within these smaller magma chambers could still separate and make modest amounts of anorthosite, but the olivine and pyroxene would remain within the crust. In other words, the overall composition of the crust would still be gabbro. A lunar crust constructed through serial

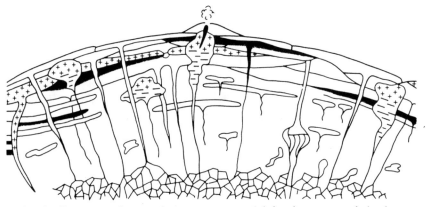

A sketch illustrating the serial magmatism model for formation of the lunar crust. Basaltic magmas invaded the crust at various times, crystallizing to form gabbro with only modest amounts of anorthosite. Although this model requires a great deal of magma, all of it does not have to be present at the same time. *From D. Walker, Journal of Geophysical Research 88 (1983): B17, copyright American Geophysical Union, with permission.*

magmatism would in no case contain more than about half feldspar.

Several ingenious experiments have been devised to estimate the average feldspar content of the highlands and so decide between the magma ocean and serial magmatism models. These estimates were based on measurements made by orbiting spacecraft, the *Apollo 15* and *16* service modules that remained in lunar orbit while the landers descended to the surface. Two kinds of experiments were done. One measured X rays produced when the Sun's rays impinged on the lunar surface. Because the Moon has no absorbing atmosphere, solar radiation is able to reach the surface directly. This irradiation excites atoms on the lunar surface to emit fluorescent X rays, and their measurement from orbit provided information on the elemental composition of the soil. A second experiment measured gamma rays, which are given off spontaneously during the decay of radioactive elements in surface materials. Other, nonradioactive elements in the soil also give off gamma rays as stray cosmic rays impinge on them, and the abundances of these elements also were estimated. The area of the Moon covered by these two experiments was of course limited to the orbital tracks of the service modules. Gamma rays were recorded continuously, but X rays could be measured only on sunlit areas, so coverage was actually limited to about 9 percent of the lunar surface area.

The orbital X-ray data gave information on the relative amounts of aluminum and silicon in surface soil at various locations. The detected ratio of aluminum to silicon was much higher whenever the orbiter traversed over highlands areas than over maria. Because most aluminum in rocks is sited in feldspars, this measurement indicated that ancient crustal rocks everywhere on the Moon's surface must be feldspar-rich, more like anorthosite than gabbro. The best estimate of the proportion of feldspar in the highlands using this method is about 75 percent. In a similar manner, the orbital gamma-ray data have been used to estimate the feldspar content of the lunar surface. Although the elements in feldspars were not measured directly using this technique, lunar anorthosite has a chemical signature that can be distinguished using gamma-ray measurements. Anorthosites contain characteristic amounts of iron, titanium, and thorium, which can be estimated from the gamma rays they exude, and orbital data on these elements have been used to corroborate the estimate of the highlands feldspar content from X-ray measurements. Furthermore, the ejecta

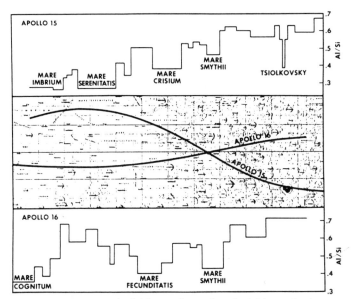

The ratios of aluminum (symbol Al) to silicon (symbol Si) on the lunar surface are shown for two orbiter tracks flown during the *Apollo 15* and *16* missions. The amounts of these elements were determined from X-ray and gamma-ray measurements onboard the orbiting spacecraft and were checked by comparing the measurements with analyzed soil samples collected at overflown landing sites. Areas of ancient highlands crust have consistently higher aluminum contents than maria. Because feldspar is rich in aluminum, this experiment supports the idea that the lunar crust is mostly anorthosite. (*NASA Johnson Space Center.*)

blankets in and around large craters were analyzed carefully, with the expectation that they would contain materials excavated from deeper regions of the crust. Although these areas tend to have somewhat higher proportions of pyroxene and olivine, their high feldspar contents demonstrate that a feldspathic crust extends to appreciable depth.

Both sets of orbital experiments appear to point to an ancient rocky crust that has a composition too rich in feldspars to have formed by serial magmatism, but one that could have formed from a global magma ocean. The estimated thickness of the anorthosite crustal layer has been revised downward from Wood's original estimate of twenty-five kilometers to about twenty kilometers, but this new figure still implies a deep magma ocean, hundreds to perhaps as much as 1,000 kilometers thick.

We know that the magma ocean must have formed and then solidified very early in the Moon's history, but these are difficult events to date precisely. The highlands rocks should record the time of their own crystallization, but it is difficult to determine their ages. Accurate ages require the isotopic analysis of several different minerals in the same rock, and anorthosites have very little of any mineral other than feldspar. A further complication is that most of these rocks have suffered numerous impacts that have reset their radioactive timepieces. The best available radiometric ages suggest that the formation of the solidified lunar crust was probably complete by about 4.4 billion years ago.

MAGMA OCEANS ON THE EARTH AND OTHER PLANETS

Specifying an adequate heat source for producing the magma ocean has always been somewhat of a problem. Previous explanations have appealed to rapid accretion of the Moon from impacting meteors, the decay of short-lived radioactive isotopes, or the heat given off during core formation. This problem has now been resolved to many planetary scientists' satisfaction. If we accept that the Moon was formed during a catastrophic impact of a Mars-size body with the Earth, we need no other heat source. Such an impact imparted a great deal of heat energy to the matter that ultimately formed the Moon, and the existence of a magma ocean seems certain if the Moon formed in this way.

What is perhaps more intruiguing is that the same giant impact also would have caused a substantial fraction, or possibly all, of the Earth to melt as well. Before the idea of a lunar magma ocean was proposed, geologists had never even seriously entertained the possibility of a terrestrial magma ocean. The reason for this omission is obvious: There were no hints, because rocks dating from the Earth's earliest history have never been found. The oldest known terrestrial rocks, found in northwestern Canada, formed a little more recently than 4 billion years ago. The interval from this time back to the planet's formation is missing from the geologic record. If a magma ocean existed on the Earth, it should have been molten at about the same time as the lunar magma ocean, that is, prior to 4.4 billion years ago.

Although the giant impact hypothesis suggests that a terrestrial magma ocean was likely, some worrisome problems with this idea have yet to be resolved. The composition of the Earth's mantle presumably holds clues as to whether the planet was ever substantially molten, just as the lunar mantle does. If melting occurred on so grand a scale as a magma ocean, crystals of olivine and pyroxene probably would have separated from the liquid, just as they did in the lunar magma ocean. Some geologists have sought the chemical imprints that such a process would impose on mantle samples, a terrestrial equivalent of the complementary europium abundances in the lunar crust and mantle. The terrestrial imprints are more difficult to interpret, however, because the extreme pressures in the Earth's interior affected the behavior of elements in ways that we scarcely understand. Some experiments have been carried out to try to determine whether elements such as nickel and cobalt are preferentially incorporated into or rejected by crystals in a magma at very high pressures. The experimentalists then predict the composition of a mantle formed by the separation of crystals in a magma ocean. Comparison of the predictions with chemical analyses of terrestrial mantle rocks brought up during volcanic eruptions suggests that crystals have not been removed, implying that the Earth was never melted to any great degree. These high-pressure experiments are very tricky, however, and there are some conflicting results. The smoking gun—chemical evidence for a terrestrial magma ocean—has eluded investigators thus far, but the process remains an intriguing possibility.

What about the other terrestrial planets? Quite plausibly, they too could have been awash with magma early in their histories, if they also suffered large impacts. George Wetherill of the Carnegie Institution in Washington has used computer models to simulate the assembly of planets from smaller protoplanets. During the late stages of the accretion process, these protoplanets become very large. In the course of thirteen separate simulations of the Earth's formation, Wetherill found that the growing planet was struck by seventeen objects at least as large as Mars. The formation of Mercury, Venus, and Mars probably would have been equally eventful, offering ample opportunities for substantial or total melting of these planets. Mercury is quite similar in many respects to the Moon, and ancient crust similar to the lunar highlands is visible in photographs of its surface. Venus and Mars, however, have had protracted geologic histories that may have erased the rocky evidence of former magma oceans.

The emerging picture of planets drowned in molten silicate seas may seem surprising, if not disconcerting, to some. Certainly the early Moon was a hellish place, a scalding cauldron of magma. Gradually this magma world crusted over, but all the while incoming projectiles opened gaping wounds and exposed the molten interior anew. This was an ocean in turmoil. By studying small bits and pieces of the lunar highlands, we are in effect standing on the shoreline, peering at this phenomenal sight. It is unproved, but probable, that the newly formed Earth and its neighboring planets also were engulfed in seas of magma.

And all this derives ultimately from conjectures made from the observation of a few bits of feldspar in lunar soil. As Mark Twain remarked in *Life on the Mississippi*: "There is something fascinating about science. One gets such wholesale returns of conjecture out of such a trifling investment of fact."

Some Suggestions for Further Reading

Heiken G. H., Vaniman, D. T., and French, B. M., eds. 1991. *Lunar Sourcebook: A User's Guide to the Moon*. Cambridge, England: Cambridge University Press. An up-to-date treatise on the Moon; Chapter 2 traces the evolution of the magma ocean concept, and Chapter 6 gives thorough descriptions and references for lunar highlands samples.

Taylor, S. R. 1982. *Planetary Science: A Lunar Perspective*. Houston, TX: Lunar and Planetary Institute. A superb book by the dean of lunar science. The book describes highlands rocks, evidence for the lunar magma ocean and for similar magma seas on the terrestrial planets.

Walker, D. 1983. "Lunar and Terrestrial Crust Formation." *Proceedings of the Fourteenth Lunar and Planetary Science Conference, Journal of Geophysical Research*, vol. 88, pp. B17–25. A technical but very readable account of the history of thought on the formation of the lunar crust.

Wilhelms, D. E. 1987. *The Geologic History of the Moon*. U.S. Geological Survey Professional Paper 1348. A thorough and well-illustrated account of the evolution of the Moon, including its magma ocean.

Wood, J. A., Dickey, J. S., Marvin, U. B., and Powell, B. N. 1970. "Lunar Anorthosites and a Geophysical Model of the Moon." *Proceedings of the Apollo 11 Lunar Science Conference*. New York: Pergamon, pp. 965–988. The paper that began this story, excellent reading to put the concept of a lunar magma ocean in perspective.

A Goddess Unveiled

Magellan *Reveals the Surface of Venus*

My parents had three children—two sons and a daughter. My siblings and I share the same gene pool, but you would never know it from our career choices. My sister is a musician, my brother is a tennis pro, and I am a scientist. Although I like to fancy myself as being musically knowledgeable and a fair doubles player, the reality is that I am probably neither, and my sister and brother likewise have little aptitude for science. We three are a lot alike, and at the same time are as different as night and day.

So too are Venus and Earth, sometimes described as sibling worlds. Both planets were constructed basically of the same matter at about the same time, and have similar sizes, masses, and densities. But until very recently, we Earthlings knew almost nothing about the surface appearance of our sister planet, even though she lives just next door. Viewed from afar as the morning or evening star, she has often been compared to a shimmering goddess, at various times called Ishtar by the Babylonians, Aphrodite by the Greeks, or Venus by the Romans. The stock and trade of this goddess are matchless beauty and love fulfilled. In more recent cultures, the name Venus remains synonymous with these themes, as illustrated by Michelangelo's exquisite Venus de Milo or Frankie Avalon's sixties' hit love song "Venus." New planetary observations, however, reveal that up close this goddess is more like an old floozy, her roughened and pocked complexion hidden under caked-on makeup of thick, creamy yellow clouds. It is the high reflectivity of this obscuring cloud bank that accounts for the planet's dazzling brightness in the night sky.

EARLY IDEAS ABOUT VENUS

There has been no shortage of scientific conjecture about the appearance of Venus under her swirling yellow veil. The obscuring mists have invoked fantasies of a hazy, swampy world reminiscent of the Mesozoic Earth, populated with lush vegetation and even foraging dinosaurs. Spectral studies, however, showed not even a hint of water vapor; instead, the clouds were found to be composed mostly of carbon dioxide. Other planetary researchers then surmised that the pervasive carbon dioxide blanket had formed by reaction of Venus' water with a vast reservoir of gooey hydrocarbons, implying the existence of a globe-encircling oil slick of nightmarish proportions. Still others argued that water vapor lines were absent from the venusian atmospheric spectrum because surface temperatures were cold enough to condense all of the water as liquid, which combined with the carbon dioxide in the atmosphere to make seas of carbonated seltzer water. Later temperature measurements showed the surface to be hotter than a self-cleaning oven, causing the oil slick and fizzy ocean hypotheses to be supplanted by the notion of a hellish planetary greenhouse run amok.

RADAR IMAGERY

The true face of Venus was finally revealed when radar measurements pierced the obscuring caul of carbon dioxide and sulfuric acid vapor. Several decades ago the first goddess voyeurs, who prefer to be called radio astronomers, snatched murky glimpses of the planet's surface by beaming powerful radar signals at Venus and catching their reflections in a huge dish at Arecibo, Puerto Rico. The variations in times required for these reflected signals to return to Earth were then converted into planetary surface elevations. The result was a picture of surface relief, really more akin to the likeness in a carved cameo than in a photograph. The resolution of individual features was not very good though, because the distance that the signals had to travel was so great compared to the height differences of surface features. What was needed for a better view was a radar platform much closer to Venus, an orbiting peeping Tom with its nose pressed against the atmospheric window.

The first spacefaring paparazzo photographer was *Pioneer-Venus*, which reached the planet in late 1978 and immediately began bouncing radar signals off the surface. The picture of Venus that emerged from the *Pioneer* data and from the earlier Arecibo results was of a planet with high-standing "continents" surrounded by low-lying "basins," something like the Earth if its oceans had been miraculously emptied of water. But what do descriptive terms such as continents and basins mean on a planet with no water? It is easy to understand the meaning of these terms on the Earth, because we relate elevations

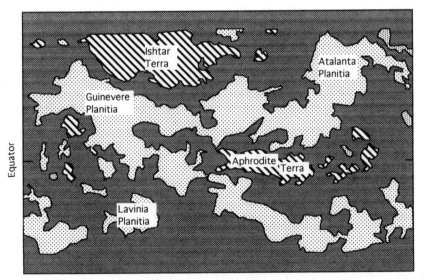

A sketch map of the surface of Venus, showing areas with elevations above the mean planetary radius (labeled terra) and below that level (labeled planitia). Prominent features are mostly named for mythical female characters.

to sea level. In the absence of oceans on Venus, however, we need a different reference level, and the one that has been chosen for global radar maps is the mean planetary radius. Specific regions, then, are described as lying either above or below the average elevation of the entire planetary surface.

Venusian highlands, defined as those areas that have elevations well above the mean planet radius, are called terrae, as on the Moon. The term planitiae, meaning "plains," is given to the intervening basinal regions. Other regions of rolling plains lie at elevations near the mean planet radius. By convention, the major features on Venus are appropriately named for females, both real and mythical. Circular, bowlike forms are named for notable historical women, as in the craters Dickinson (Emily) and Earhart (Amelia), and other features bear names of mythological goddesses and heroines, as in Aphrodite terrae and Guinevere planitiae. A glaring exception is Maxwell Montes, a great mountain chain that honors a male physicist whose work on radio waves made the discovery of radar possible. Two other small highlands regions, Alpha and Beta, have genderless names that were grandfathered in when the planet's nomenclature turned feminist.

Beginning in 1983, Soviet *Venera* 15 and 16 orbiting spacecraft provided new radar images with even higher resolution, delineating venusian features as small as a kilometer across. But the most spectacular views of the goddess have been provided by *Magellan*, an American spacecraft inserted into a near-polar venusian orbit in 1990. *Magellan* has mapped virtually the entire surface, and its improved radar system has resolved blemishes almost as small as a football field.

Most radar systems send out one signal at a time and process each echo by itself. The detail seen with conventional radar systems depends on the size of the antenna that receives the echo. Because a large antenna mounted on a small spacecraft such as *Magellan* would be difficult to maneuver, its designers employed a special technique to improve resolution. *Magellan* emits several thousand radar pulses each second, and fast computers accumulate many echoes at once. At the end of each orbit, the spacecraft then transmits these data back to Earth, where they are processed into images. In this manner, *Magellan's* small antenna produces images that would be comparable to those of much larger radar receivers.

Venus sat remarkably still for her unsanctioned *Magellan* portrait, but not perfectly motionless. The planet rotates very slowly, requiring 243 Earth days to make one complete turn, corresponding to one venusian day. A *Magellan* mapping cycle was designed to take one full day as reckoned on Venus. As the planet slowly rotated, every place on the entire globe eventually inched its way underneath *Magellan's* orbital position. To radar beams, the goddess was naked as a jaybird, with the spacecraft poised to expose her every flaw. During the first mapping cycle the spacecraft made 1,789 orbits, continuously recording radar images of the swatches of terrain below its flight path. The resulting pictures consisted of long, visual "noodles," representing strips of ground 20 kilometers wide and 17,000 kilometers long. The photographic strips from adjacent orbits, which overlapped slightly, were ultimately laced together to make photomosaics.

The composite radar images look superficially like shadowed black-and-white aerial photographs taken in visible light, but the similarity can be deceptive. Features that appear dark in radar photographs might actually appear bright to our eyes. This difference arises because radar signals do not interact with solid matter in exactly the same way that light rays do. Only part of the radar echo that the spacecraft antenna senses is actually reflected from the surface; the rest of the signal penetrates a short distance into the surface and is scattered back. The angle at which the radar beam strikes the surface also critically affects the radar return, so that the same area will appear different when the viewing angle changes. This can be a problem even for the experts—the *Magellan* science team had to retract an announcement of the discovery of crustal movement between the first and second mapping cycles, when they found that a radar-imaged "landslide" was actually an artifact of different viewing angles. Radar also is sensitive to the manner in which the surface materials conduct electrical currents, so the presence of conductive minerals may make the surface more reflective. All of these factors combine to make the geological interpretation of radar images ambiguous at times. I suppose you might say that the differences between radar and visual images accord Venus some modesty, but in reality it is precious little.

The sheer quantity of information produced by *Magellan* is staggering. During the first mapping cycle, the spacecraft sent to Earth 800 million bits of radar data, covering most of the venusian surface. It is probably fair to say that the surface of Venus is now probably better characterized than that of our own planet, because features over much of the Earth's surface are discretely cloaked by water. Faced with the problem of having far more images than could be reasonably published in this book, I have decided to concentrate on a few disfiguring blemishes on the goddess that provide particularly interesting geologic perspectives.

TECTONISM

Planets, like babies, are born with unwrinkled skin and become gnarled with age. The Earth's most prominent wrinkles are towering mountain ranges such as the Rockies, Alps, Himalayas, and Urals, as well as sinuous midocean ridges that snake along the ocean floors. These features are the direct result of the movements of large crustal segments, or plates. The term tectonics is often used by geologists to note the structural changes in large masses of rock. Plate tectonics describes how huge, segmented slabs of rigid crust move about and jostle each other, sometimes crunching together and at other times pulling apart. Plate collisions crumple, fracture, and uplift the crust, producing the elongated belts of buckled and faulted rock that we recognize as mountain chains. Midocean ridges occur where plates pull apart, creating yawning fracture systems that swell with magma from below.

What drives the Earth's restless plate motions is internal heat, rising through the mantle in great plumes that bring hot rock and magma toward the surface. In other areas, the opposite process occurs, as cold crustal rocks are dragged down into the mantle and recycled. The rigid crust perched above these roiling cells breaks into plates that are pushed away from areas of upwelling and dragged toward areas of downwelling. The surfaces of the Moon, Mercury, and Mars have no plates, but these worlds are very different from the Earth, both in terms of size and the amount of heat in their interiors. Because of its size, the Earth retains enough internal heat to create the convection cells that power plate movements. We might predict that plate tectonics also would occur on its similar-size sibling.

The unsightly wrinkles on Venus are every bit as noticeable as their terrestrial counterparts, but they are not nearly so common.

Wrinkled crust occurs mostly on terrae. It often forms complex patterns of intersecting ridges and valleys called tesserae, meaning "tiles," a reference to the polygonal aspect of these regions when viewed from great distances. Several episodes of tectonic movement must have occurred to produce intertwined folds with different orientations.

These leathery wrinkles are clearly tectonic structures, but they apparently did not form during collisions of large rigid plates. From geologic studies of the Earth, we know that sites of plate convergence are marked not only by elongated mountain belts but also by chains of volcanoes and deep trenches, both the result of one plate diving beneath the other. Volcanic chains and deep trenches are not associated with venusian mountain belts. If plates had collided to make the tesserae, elsewhere on Venus these same plates must have pulled

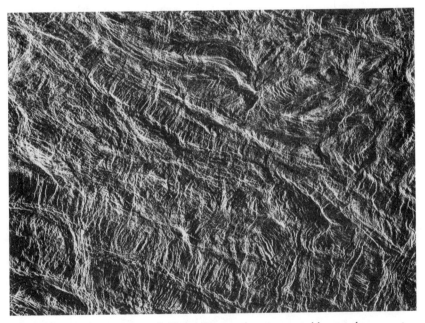

Radar image of a portion of Alpha Regio, showing wrinkles in the venusian crust. This complex mountainous region has experienced several periods of compression, producing long, northwesterly trending structures that intersect shorter, northeasterly ridges and valleys at nearly right angles. The long dimension of this photomosaic measures 150 kilometers, about the width of the Appalachian chain. *(Jet Propulsion Laboratory.)*

apart. On Earth extension causes collapse, and magmas well up and fill the intervening rifts between the separating plates. Although extensional structures do occur on Venus, they are spread over vast areas of the landscape rather than being localized at plate boundaries. Clearly, the venusian crust has been crumpled and shortened to make wrinkles, but we could not call this process "plate" tectonics. The surface of Venus is effectively one gigantic plate that has not broken apart with age, as has the Earth's crust.

VOLCANISM

This difference in crustal behavior certainly cannot be attributed to lack of internal heat and convection on Venus. The simmering interior of the goddess is clearly revealed by ample blisters and pimples on her surface. In fact, Venus is literally encrusted with thousands of volcanic warts, all testifying to the fact that the planet's interior is hot and that mantle convection occurs. These disfiguring protuberances take many bizarre forms, and we will consider only a few.

Parts of the face of Venus are decorated with circular pimples, euphemistically called domes by planetary geologists. The shapes of the domes are somewhat like pancakes, typically measuring about twenty-five kilometers in diameter with heights of a kilometer or so. Their tops contain numerous deep pits whose bottoms lie nearly at the level of the surrounding plain, so perhaps the shapes of these features might be even better characterized as English muffins. Much smaller domes with similar shapes occur on Earth, where sticky lava with the consistency of toothpaste erupts and hardens into rhyolite rock or glassy obsidian, the volcanic equivalents of granite. Complex fracture patterns that decorate the tops of the domes may have formed when already congealed lava was stretched upward as additional pulses of magma extruded from below. Alternatively, molten material in the interiors of the crusted-over domes may have withdrawn to deeper levels, causing collapse and fracturing of the overlying solidified lava. Bright rims around the domes in radar images indicate the occurrence of debris from rockslides on the slopes of the domes.

The pimples are distinctive, but the most noticeable volcanic disfigurements of Venus resemble popped blisters. These features are

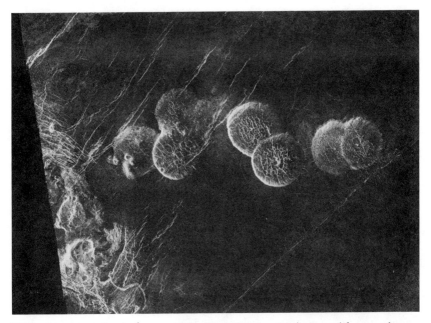

These seven circular domes on Alpha Regio have central pits and fractured tops. Each dome is about twenty-five kilometers in diameter. They are thought to be solidified eruptions of sticky lava. Some fractures on the plains are covered by the domes, whereas others slice through them, indicating that fracturing occurred both before and after the eruptions. (*Jet Propulsion Laboratory.*)

large circular structures, ranging in size up to hundreds of kilometers across. Called coronae by geologists, these circles are defined by concentric ridges and moats commonly filled with lava. Many coronae are surrounded by radial networks of varicose veins, interconnected fractures that resemble spider webs. For obvious reasons, Soviet researchers have christened these webs and their associated coronae arachnoids. These curious features apparently were formed by ascending magmas that did not break the surface but arched the overlying crust upward. The stresses induced fractures in the crust surrounding the coronae. At some later time, the magmas supporting many of the blisters withdrew, so they collapsed into circular depressions, like a fallen soufflé.

The coronae indicate that very large magma chambers existed at one time in the subsurface. Such chambers on the Earth, once congealed and unroofed by erosion, are called plutons. But our plutons are normally much smaller than those that must have made the coronae on Venus. Why venusian magmas should collect into such huge

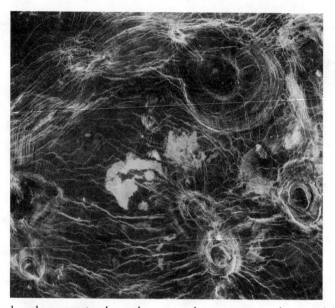

This radar photomosaic shows large circular coronae, each several hundred kilometers across, and numerous fractures, both produced by magma upwelling from below. These features are called arachnoids because of their resemblance to spiders on a web. The bright patches near the center of the figure are lava flows. This area of Bereghinya Planitia is approximately 1,600 kilometers square, about the size of Colorado. *(Jet Propulsion Laboratory.)*

blobs remains a mystery, but this produces a distinctive deformity that is widespread on the surface of the planet.

Venus also displays irregularly shaped, radar-bright birthmarks, the traces of lavas flowing on the surface. Some of these volcanic flows extend for thousands of kilometers, much longer than comparable features on the Earth. They also tend to be thinner with gentler slopes. The lava that composed these flows must have been extremely fluid, much different from the sticky toothpaste that formed the domes. Perhaps the broiling surface temperature kept these lavas from cooling too quickly. The flows sometimes can be traced back to source vents on the sides of volcanic mountains.

Domes, coronae and arachnoids, lava flows, and myriads of other volcanic features all indicate that Venus' interior was, and probably still is, awfully hot. Volcanic bumps and sores are widespread but are not localized at plate boundaries, as they are on the Earth. Some planetary geologists have coined the term blob tectonics to describe the crustal architecture of Venus. What they mean is that heat continually rises to the surface at certain localities, "hot spots" if you will, causing melting of the mantle and generating clusters of large subsurface plutons or eruptive centers. These hot spots are like blowtorches, burning into the crust from below. The Earth too has hot spots. Hawaii, for example, is a chain of volcanic islands etched into the overlying Pacific plate as it creeps over a stationary plume. But Venus has no restless plate motions to smear the surface expression of the hot spots. If Venus has a hot, convecting interior, why are there no plates?

One idea is that the frightfully hot Venus surface temperature softens the crust to greater depth than on Earth. This warming might make the crust behave like chewing gum and so prevent it from breaking into plates. If the crust and the underlying mantle are similarly pliable, they may behave as if they were coupled together. This is a very different situation from what exists on the Earth, where a partially melted transition zone separates the cooler, rigid plates from the hot, convecting mantle below, allowing plumes to break the crust into plates and drive them around the planet. If Venus has no such boundary, convection may extend through its single-plate crust nearly to the surface, as manifested in the volcanic warts and blisters so common in radar images.

IMPACT CRATERS

Not all of the disfigurements on Venus are due to internal processes. Let's now examine some ugly pockmarks caused by external forces. Like other planets, Venus has been pummeled by meteors, and their scars are apparent in many radar images. These craters are similar in many respects to the impact craters on other planets, with terraced inner walls and central peaks. Blankets of ejected debris surround the craters, their rough textures enhanced magnificently in radar imagery, making them appear almost like flower petals. Some craters also show outflows of melted rock generated during the impacts.

But there is also a stark difference between venusian craters and the meteor scars on other worlds. On the Moon, craters of all sizes occur, their numbers increasing exponentially as their diameters decrease. In marked contrast, Venus has very few craters smaller than

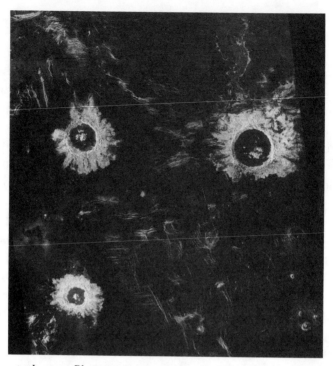

This area in Lavinia Planitia is sometimes called the crater farm. It contains three large impact craters, with diameters ranging from thirty-seven to fifty kilometers. The craters display central peaks and ejecta blankets that resemble flower petals. (*Jet Propulsion Laboratory.*)

thirty kilometers in diameter, and those that do exist are oddly clustered into tight groups. The venusian meteor pox was apparently selective. Large meteors have excavated huge craters on Venus, but meteors smaller than a kilometer or so in diameter must have been destroyed during their transit through the planet's dense atmosphere. The clusters of small craters suggest that even modest-size meteors were broken into pieces before they impacted. The atmospheric pressure at Venus' surface is ninety times that of the Earth's atmosphere, approximately the same as the crushing pressure of the ocean at a kilometer's depth. This dense blanket apparently acts as a screen for smaller meteors, the arrivals of which are marked only by crater clusters or dark splotches that are the ghostly images of their shock waves. As gauged from the proportions of small to large craters on the Moon, the venusian atmosphere has filtered out many tens of thousands of small impactors. Venus' suffocating veil apparently affords much better protection against incoming projectiles than does Earth's atmosphere.

The abundance of large craters on Venus has been used to estimate the ages of volcanic plains, although the absence of small craters limits the accuracy of the results. The ages of the planitae are everywhere the same, about 800 million years. It appears that Venus may have been substantially resurfaced, like a parking lot, by volcanism at that time. Perhaps Venus finally had enough of her ancient, ravaged visage, and opted for a face-lift. The operation was successful in erasing the older crater population, although other impacts since that time have chipped away at her resculptured face.

WIND STREAKS

If all that were not enough, *Magellan* has revealed the embarrassing fact that the goddess could use a bath. Thousands of stains, some light and some dark, cover her surface. Long, straight streaks stretch for hundreds of kilometers, cutting across fractures, coronae, and other surface features. Surface roughness determines whether these stains are light or dark to radar, with smooth areas being highly

This dark streak is a blanket of windblown dust, stretching for thousands of kilometers across the venusian surface. This particular streak is partly covered by a bright lava flow. Consistent orientations of streaks define a global wind pattern that is not apparent in cloud formations. (*Jet Propulsion Laboratory.*)

reflective or bright. The light streaks are probably wind-scoured bed-rock in the lee of obstacles, and their dark counterparts are thought to be windblown dust littering the surface. High in the atmosphere, winds blow at several hundred kilometers per hour, but down on the surface the air is so thick and soupy that the wind is no faster than walking speed. This appears to be just enough wind velocity to move sand grains. The consistent orientations of these streaks suggest that wind always blows from the same direction on Venus. The global pattern pieced together from windblown streaks indicates that winds on the surface travel from the poles toward the equator and then return aloft.

THE FACE OF THE GODDESS

From this brief assortment of radar images, it is clear that the revealed face of Venus is clearly not like our predecessors thought. She is horribly wrinkled by stress and age. She has had numerous serious outbreaks of volcanic acne, which have left her pimpled and blistered, harboring lava birthmarks and varicose veins. She has been pock-marked by large meteors continuously, despite valiant efforts at self-protection. An attempt to polish her image with a volcanic face-lift was reasonably successful, but many scars have been added since that operation. Her face is dirty, stained by long streaks of windborne dust. She runs a constant fever and has no water to slake her thirst or soften her rough edges. It is no wonder that she hides herself behind a veil of clouds.

But are we expecting too much of the goddess of love and beauty? She is, after all, the Earth's sibling, and our planet is no beauty pageant contestant either. Come to think of it, perhaps these two sisters look just the way we would want them to. Age has given them character; experience and adversity have made them geologically interesting. Looking carefully at their worn and ravaged faces, we can decipher the processes that have molded and shaped them. Venus, naked to radar, can no longer keep her secrets hidden, and now at last we have begun to learn about her long and interesting geologic history. What would be duller, or sadder, than a four–and–a–half–billion–year–old lady with no story to tell?

Some Suggestions for Further Reading

Burgess, E. 1985. *Venus: An Errant Twin*. New York: Columbia University Press. A nice, nontechnical summary of the history of Venus exploration through *Pioneer-Venus*.

Head, J. W., and Crumpler, L. S. 1990. "Venus Geology and Tectonics: Hotspot and Crustal Spreading Models and Questions for the Magellan Mission." *Nature*, vol. 346, pp. 525–533. A technical paper outlining what was known about Venus before the *Magellan* encounter.

Journal of Geophysical Research (Planets), vol. 97, no. E8 (1992). A thick issue devoted to *Magellan* results. These technically difficult papers encompass most of what is now known about Venus from this important mission. Especially relevant papers are: "Magellan Mission Summary," by R. S. Saunders et al.; "Venus Volcanism," by J. W. Head et al.; "Venus Tectonics," by S. C. Solomon et al.; "Geology and Distribution of Impact Craters on Venus," by G. G. Schaber et al.; and "Aeolian Features on Venus," by R. Greeley et al.

Science, vol. 252 (1991). This issue also contains many excellent technical articles by the *Magellan* science team. Especially relevant to this chapter are back-to-back papers on the following topics: "Venus Volcanism," by J. W. Head et al.; "Venus Craters," by R. J. Phillips et al.; and "Venus Tectonics," by S. C. Solomon et al.

At the time of writing of this book, the *Magellan* data were relatively new and few popular articles were available. The following selections contain excellent *Magellan* photographs with descriptive captions, but offer little insight into global geologic processes on Venus:

Beatty, J. K. 1991. "Venus in the Radar Spotlight." *Sky & Telescope*, vol. 82, no. 1, pp. 24–30.

Burnham, R. 1991. "Update on Magellan." *Astronomy*, vol. 19, no. 2, pp. 44–46.

Eicher, D. J. 1991. "Magellan Scores at Venus." *Astronomy*, vol. 19, no. 1, pp. 34–42.

Twinkle, Twinkle, Little Lump

———

Abundance of Elements in the Solar System

At first glance, the blazing centerpiece of our solar system would seem to have little, if anything, in common with a lump of clay. It is nonetheless a fact that much of what we know about the composition of the Sun has been gleaned from a dessicated mud puddle. Before we learn about the mud puddle, let's take a look at some characteristics of our local star.

The Sun cannot, by any stretch of the imagination, be treated as a geologic object, and thus it may seem an inappropriate topic for this book. It dwarfs the rest of the solar system, containing more than 99 percent of its total mass. Its incandescent surface decorated with magnetic arches and its fusion-driven interior furnace have no geological counterparts, and nowhere within this furiously burning orb is anything that remotely could be likened to rocks or minerals. On an atomic level, however, the Sun is composed of the same ninety-two elements that comprise the planets, though in somewhat different proportions and physical states. The problem we will consider, then, is how to determine the elemental composition of the Sun, which is tantamount to the average chemical composition of the entire solar system. This composition is sometimes referred to as the "cosmic" abundance of the elements. As you will see shortly, deriving the cosmic abundance table is partly a geologic exercise, in that it involves the analysis of rocks.

METEORITES AS PROXIES FOR THE SUN

The quest for an accurate tally of the elemental building blocks comprising our surroundings has been going on now for more than a century. In 1889 geochemist Frank Clarke of the United States Geological Survey read a paper before the Philosophical Society of Washington entitled "The Relative Abundance of the Chemical Elements." His was the first of numerous lectures and articles by many authors attacking this subject. Although Clarke's paper was an admirable attempt to deduce the composition of the cosmos, it was fatally flawed. He incorrectly presumed that the chemical composition of the Earth's crust was the same as the whole planet, which in turn accurately reflected the composition of the solar system. We know now, however, that the Earth has been segregated into a core, mantle, and crust with radically different compositions, so that there is no longer anyplace on or within our planet where one can obtain a sample that has the composition of the whole body.

A rather different approach was suggested in 1901 by Oliver Farrington of the Smithsonian Institution. Farrington noted that the most abundant minerals in meteorites comprise only an insignificant part of terrestrial crustal rocks, but he suggested that this distinction might disappear if scientists could compare the composition of meteorites to that of the entire Earth. This novel idea set the stage for using meteorites as a basis for estimating our planet's bulk composition and, by extension, for inferring the cosmic abundances of elements.

The first person to take Farrington's idea and run with it was a physicist, William Harkins of the University of Chicago. The early part of the twentieth century was a time of great activity in atomic physics, culminating in the modern view of the atom with its tiny nucleus surrounded by dizzily spinning electrons. Electron clouds from adjacent atoms mesh to bond them together into molecules, so electrons are the bread and butter of the chemist. But an element's identity, its soul, is sited in its nucleus, and Harkins busied himself developing theories of the structure of the nucleus based on its rudimentary constituents—protons and neutrons. (At that time, further subdivision of matter into quarks, hadrons, and such was unthinkable.) In 1917 Harkins concluded that nuclei with odd numbers of protons, odd *atomic numbers* in the lingo of physics, were less stable than those with even atomic numbers. He felt that this difference in atom stability should translate into different relative abundances of the elements. Accepting Farrington's argument that the compositions of meteorites were more representative of cosmic abundances than were terrestrial rocks, Harkins used meteorite analyses to test his prediction. Sure enough, he found that elements with even atomic numbers were more abundant than those with odd atomic numbers, an observation now appropriately known as Harkins' rule. In fact, the even-numbered elements formed a lopsided 98 percent of meteoritic matter. Harkins' work, grounded in the chemical compositions of meteorites, clearly showed that it was the nuclear properties of elements, not the configuration of electrons spinning about the nuclei, that determined their abundances.

In defining element abundances, Harkins used a weighted average of the analyses for meteoritic stones and irons compiled by Farrington. Over the years, this preferred meteorite composition has been refined continually by other chemists who analyze meteorites, and who have gradually come to be known as cosmochemists. By judicious selection of the least altered meteorite samples and with the addition of elements occurring in minute quantities analyzed by ever more sophisticated techniques, the cosmic abundance table gradually has become more accurate and more complete.

ANALYZING THE CHEMISTRY OF THE SUN

Although Harkins inferred that his cosmic abundance table provided an estimate of the chemical composition of the solar system and hence of the Sun, he had no way of testing this conclusion. But help, in the form of a procedure for directly measuring the chemistry of sunbeams, was on the way. Of course, sunbeams are merely the benign outward manifestation of a hellish nuclear furnace located deep in the Sun's interior. There, atoms of hydrogen are stir-fried, fused together to make helium atoms that weigh ever so slightly less than the original hydrogen. The difference, 0.7 percent of annihilated mass, represents a substantial amount of newly created energy that slowly works its way outward toward the surface, finally to escape as light and heat. And herein lies a problem in determining the solar composition. The energy created in the Sun's center must travel through a great swath of stellar matter, about which we know very little, before we can measure it. The temperatures, densities, and mechanisms of energy transfer in the solar interior only can be guessed at for the most part, leading to considerable uncertainties in the interpretation of the light and heat energy that is finally emitted.

To estimate the Sun's chemistry from this released energy, astronomers employ the technique of spectrographic analysis. A *spectrograph* is a kind of prism, a device that splits sunlight into its component colors. The prism phenomenon is familiar to most people, even though the spectrograph may not be. Most astronomical spectrographs use finely ruled gratings rather than glass prisms to disperse the light, but the result is the same. Sunlight dispersed into a spectrum is recorded on a long piece of photographic film, producing a succession of dark vertical lines. Viewed in this way, the solar spectrum does not resemble a rainbow at all, even though it encompasses every color the eye can see, from violet to red. This range of color is caused by light vibrating at different wavelengths.

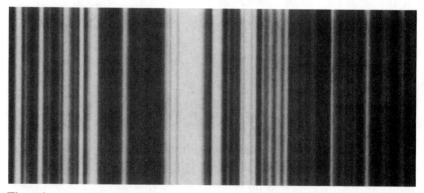

The solar spectrum contains dark absorption bands that can be used to estimate the Sun's chemical composition. This figure shows only a small part of the solar spectrum. A given element produces more than one absorption line. (*Mount Wilson Observatory.*)

The dark bands in the solar spectrum are called Fraunhofer lines, after a nineteenth-century German physicist who painstakingly cataloged hundreds of them. Each particular element, when it is in gaseous form, produces its own characteristic pattern of Fraunhofer lines. As the energy produced in the Sun's interior emerges, it must pass through a cooler gaseous envelope. The result is an electron tango that produces absorptions of energy at certain wavelengths and a series of dark absorption lines. Fraunhofer lines do not come just one to a customer, but a given element can produce many such lines, each caused by electrons absorbing energy of varying wavelength and moving to different places on the dance floor.

When the astronomical spectrograph was invented, it became possible from a remote vantage point on the Earth to identify the elements that comprised the photosphere, the bright, gaseous surface of the Sun. One new element was even discovered using this technique, when Norman Lockyer in 1868 stumbled across an absorption line in the solar spectrum that he could not identify. The Greek word for "Sun" is *helios*, from which Lockyer derived *helium*—the name he gave to this new element. Not until years later was a gas discovered on Earth that produced the same spectral line that Lockyer had discovered in the Sun.

The origin of solar Fraunhofer lines, and even the identity of some of the elements to which specific lines correspond, have been understood since the middle of the last century. But solar spectroscopy was practiced as a qualitative science, merely using the existence of lines to catalog which elements were there. Spectroscopists knew that the greater the concentration of a particular element, the stronger its absorption of light at certain wavelengths should be, and they routinely used spectrographs to measure element abundances in laboratory samples. Converting Fraunhofer lines in the solar spectrum into element abundances, though, was another matter entirely. You see, the darkness of the absorption bands depends not only on the concentration of the element in the Sun, but also on other factors such as temperature and pressure. In 1927 Princeton astronomer Henry Russell discovered a way to quantify the solar spectrum, so as to determine directly the chemical composition of the Sun. Russell was a superb mathematician, and without the assistance of a computer he was able to carry out the complex calculations necessary to calibrate solar

Fraunhofer lines. He then used his own eyeball estimates of the shadings of the dark lines to derive the solar abundances of seventeen elements. Although Russell's approach was crude by today's standards, his results were remarkably accurate. He found that the Sun consists almost entirely of two elements—hydrogen, which makes up about 90 percent of its mass, and helium, comprising about 9 percent. Carbon, oxygen, and nitrogen are several thousand times less abundant, and the other, heavier elements are present in even lower concentrations.

Since Russell's pioneering contribution was published, many other astronomers have employed spectrographic techniques to reveal more details of the chemistry encoded in shafts of sunlight as well as the light of other stars. Russell's theoretical corrections that made possible the transformation of solar Fraunhofer lines into element abundances have been improved upon by experimental determinations, in a few cases resulting in significant revisions in our estimate of the Sun's chemistry. For example, Russell's measurement of the amount of iron in the Sun was a matter of contention for many years. The original correction for iron Fraunhofer lines was found to be incorrect, because of a trivial error in the experimental procedure used to derive it, and the cosmic abundance of this element now has been increased fourfold.

Although spectroscopy has yielded a lot of information about solar composition, the inventory of elements in the Sun is still not complete. Some elements, such as argon, boron, and arsenic, either have no visible absorption lines or their lines are obscured by the overlapping lines of other elements. Other elements are present in such minuscule quantities in the Sun that their abundances cannot be accurately measured.

METEORITES AGAIN

As if he had not already done enough, Russell also provided a way around these problems. In his 1927 article, Russell showed that the relative abundances of elements that he felt were accurately determined in the Sun agreed remarkably well with their measured abundances in meteorites. If this is true, it should be possible to fill in the gaps in the solar abundance table by analyzing those elements in meteorites! An old and familiar adage states that "A bird in the hand is worth two in the bush." In the present context, we could restate this as "It is easier to obtain an accurate and complete chemical analysis of a meteorite in the laboratory than of the Sun from afar." And that is just the procedure we use today. The current cosmic abundance table is a combination of meteorite analyses and solar spectral analyses, a shuffling of data from two sources that is justified by the chemical correspondence between these objects.

A modernized version of what Russell saw when he compared his solar composition to meteorites is shown in the graph. It requires some explanation. Because there is such a huge variation in the abundances of hydrogen, helium, and the other elements in the Sun, this graph uses a logarithmic scale; hence, each marked division on both the horizontal and vertical axes signifies a tenfold increase in abundance. Direct comparison of the chemistry of the gigantic Sun and a tiny meteorite sample also requires that we employ a trick in the construction of this graph. The Sun obviously contains many more atoms of any given element than does a golf ball–size meteorite. How can we compare the many trillions of cobalt atoms in the Sun to

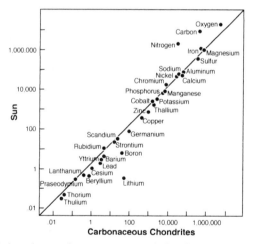

A comparison of the chemical composition of the Sun, determined from its spectrum, with analyzed meteorites, in this case carbonaceous chondrites. The atomic abundances of each element are given, relative to 1 million silicon atoms. The fact that most elements fall along the diagonal line indicates that the Sun and the meteorites have very similar compositions, except for volatile elements such as hydrogen and helium (not shown), which are much more abundant in the Sun.

perhaps only a few thousand cobalt atoms in a tiny chunk of meteorite? What we really are interested in learning is the abundance of
cobalt *relative to other elements* in both objects or, put another way, the
ratio of cobalt atoms to some other element. All the values shown in
the graph are ratios of the number of atoms of each element to a
million atoms of silicon. (Silicon is a convenient choice because its
concentration has been measured fairly accurately in the Sun and in
meteorites.)

If the chemistry of the Sun and meteorites are precisely the same,
then every element must plot along the diagonal line in this figure.
Cobalt does fall along the line, indicating that both the Sun and
meteorites contain approximately 1,000 cobalt atoms for every 1
million silicon atoms. The observation that most elements plot along
this diagonal line supports the idea that meteorites do have cosmic
compositions. If you could somehow collect a sample of the Sun and
cool it so that the gaseous atoms condensed into solid form, you
would, more or less, have made these meteorites. I say "more or less"
because most of the solar sample would consist of hydrogen and
helium, very little of which are contained in meteorites. Other elements such as carbon, nitrogen, and oxygen, so-called volatile elements that form gases even at low temperatures, also are less abundant
in meteorites than in the Sun. So meteorites can be thought of as
distilled Sun, a sort of solar sludge that lost much of its original
inventory of volatile elements. However, the condensable elements,
"metals" in the parlance of astronomers, are present in meteorites in
their original cosmic proportions relative to each other.

Besides the volatile elements, there are several others in this graph
that do not seem to be very well behaved and fall well off the cosmic
abundance line. Curiously, lithium, beryllium, and boron are more
abundant in meteorites than they are in the Sun. We believe that
these elements have been destroyed systematically within the Sun by
nuclear processes during its 4.5–billion–year history. In this regard,
meteorites record the primordial composition of the solar system
even better than does the present-day Sun. Of course, the Sun's
composition has changed in another way as well, as 700 million tons
of hydrogen have been processed into helium soot during each second
of its existence, but neither of these elements has been incorporated
into the meteorite to any great degree so we cannot see this effect.

THE CHONDRITE ORGUEIL

Actually, this Sun-meteorite comparison has not been made with just any old meteorite, but a very specific type called a Type 1 carbonaceous chondrite. Such meteorites are rich in organic matter, which, besides accounting for the name, makes them black and gives them a decidedly grungy appearance. Carbonaceous chondrites occupy a very special place in the field of cosmochemistry, because of all the known types of meteorites, they contain the highest proportion of volatile elements and consequently are most similar in composition to the Sun.

There are only a handful of Type 1 carbonaceous chondrites, and with one exception, they are all either small meteorites or mostly have been exhausted through destructive chemical analyses over many years. The lone exception fell as a shower of about a hundred separate stones over France on the evening of May 14, 1864. Twenty specimens were collected near the village of Orgueil, from which this meteorite took its name. Over the years, samples of Orgueil have been widely distributed and extensively analyzed. In fact, the entire Periodic Table has been analyzed many times over in this meteorite, and it is probably fair to say that Orgueil is now the most thoroughly characterized matter known to man, at least from a chemical point of view. Consequently, the cosmic abundance table owes a great deal to this one chondrite, arguably one of the most important meteorites in our present collections.

Because its composition so closely matches that of the Sun, the Orgueil meteorite was once considered to be made of solid matter that condensed directly from the gaseous solar nebula. This primitive matter then remained pristine and untouched, an amazing feat considering that it formed some 4.5 billion years ago. But however appealing that idea was, it has not stood the test of time. Scientists often devise clever and interesting titles for their presentations at research conferences, in the hope of piquing the curiosity of their potential audience. Of the thousands of titles I have seen, one of the most effective, for my money, is "The Search for the Holy Grail and Why Orgueil Is Not It." John Kerridge of the University of California at Los Angeles gave this talk to a packed house at a meeting of the Meteoritical Society in 1977. The subject of his talk was upsetting to many in the audience then, and it remains a somewhat contentious subject today. Kerridge's thesis was that Orgueil and other Type 1 carbonaceous chondrites are not primitive matter at all, but have been thoroughly altered from their original state.

I have been a party to this debate myself over the years, taking Kerridge's side in the argument. The evidence, at least to me, seems incontrovertible and glaringly obvious. Orgueil is cut through and through with thin white veins, consisting of the minerals epsomite (magnesium sulfate, used in epsom salts), gypsum (calcium sulfate, used in wallboard), and, less commonly, calcite (calcium carbonate, the prime ingredient of limestone). These same minerals commonly form from evaporating seawater on the Earth, so it seems likely that the Orgueil veins may have been deposited from some kind of solution, perhaps coursing through a regolith on the parent asteroid.

The bulk of Orgueil is dark material composed of a complex assortment of tiny mineral grains, most no larger than ground pepper. A prominent mineral among these grains is serpentine, a magnesium-iron silicate named from its markings and green color, which are reminiscent of a snake's skin. A second important mineral is smectite, a form of clay. Both of these minerals contain water bound into

A microscopic view of a paper-thin slice of the Orgueil meteorite, showing that it is crisscrossed with white veins of sulfate and carbonate minerals that precipitated from solutions. The black material consists mostly of fine-grained serpentine and smectite, also products of aqueous alteration.

their crystal structures. In the original interpretation, these hydrous minerals were thought to have formed in the solar nebula. However, this conclusion has been called into question by newer calculations that require extremely long periods of time for reactions between water vapor and silicate grains in the tenuous nebula gas. A better explanation is that the serpentine and smectite in Orgueil formed by reactions of other, dry minerals with liquid water on an asteroid. The water was presumably derived from melting ice that was originally incorporated into the rocky asteroid.

Orgueil clearly is an altered rock. In fact, this carbonaceous chondrite looks like nothing so much as a congealed mud puddle. It is as much cosmuck as cosmic. Although many scientists now accept this interpretation, we are left with the quandary of why lumps of dried-up mud should mimic rather precisely the chemical composition of the Sun. The original minerals of this meteorite have been replaced completely by a new set of hydrous minerals, and solutions have precipitated other minerals in veins. Mineralogic alteration of rocks by circulating water on the Earth normally wreaks havoc with rock chemistry, dissolving and carrying away some elements while importing others. To make matters worse, detailed analysis of the oxygen isotopes in Orgueil indicates that the water and the rock exchanged oxygen atoms freely. It is difficult to see how the isotopic composition of oxygen, which makes up more than half of the rock by weight, could have been changed while the proportions of other elements in much smaller quantities were not affected. I can offer no ready explanation for this puzzle of how an asteroidal mud puddle preserved the composition of its nebular forebear.

REFINED COSMIC ABUNDANCES

Since we have no answer for this particular dilemma, let us simply acknowledge that Mother Nature has left some interesting problems for future generations to solve and move on. If we accept the observation that Orgueil and the Sun apparently share the same chemistry, except for the most volatile elements, we can further refine the solar abundances, substituting Orgueil analyses wherever the solar measurements are lacking or in doubt. A modern summary of the cosmic abundances of all the naturally occurring elements, plotted in order of increasing atomic number, reinforces some of the conclusions we have already made. First, the elements vary in abundance over twelve orders of magnitude, stairstepping down from light atoms to heavy atoms. Next, the abundances of elements just to the right of hydrogen and helium are strongly depleted, and those near iron seem to be enriched. Finally, the elements form a zigzag pattern, a confirmation of Harkins' rule in which odd atomic numbers are less abundant than even.

Such a summary graph contains a goldmine of information for astrophysicists who worry about how elements are forged in stellar furnaces. The stairstep abundance pattern from light to heavy elements indicates that the elements were created from left to right, that is, the heavy elements were made from the more abundant light elements. The sawtooth pattern is a reflection of the relative stabilities

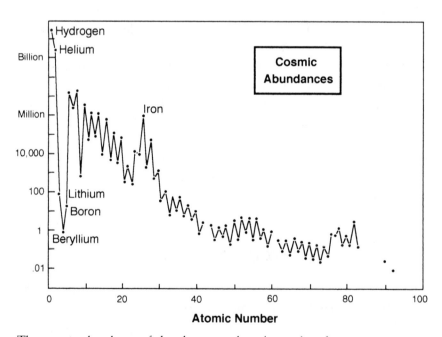

The cosmic abundance of the elements, plotted in order of increasing atomic number (the number of protons in the nucleus). All abundances are relative to 1 million silicon atoms. The way the pattern slopes downward to the right, its sawtooth appearance, and its peak at iron and valley at lithium, boron, and beryllium provide important information on how the elements were formed in stars.

of different nuclear arrangements. And the low amounts of lithium, boron, and beryllium are caused by nuclear reactions that have depleted these elements in the solar furnace over its long history. The peak represented by iron and its overly abundant neighbors reflects the fact that the nuclei of these elements are particularly stable. From information such as this, detailed models for the formation of the elements, the elegant alchemy of twinkling stars, have been constructed.

Perhaps applying the term cosmic to the composition of the Sun is inappropriate, an indication of our egocentric view of the universe. After all, there is nothing grand or even representative about our local star. The peculiar chemistry of the Sun is an accident, a random mix of interstellar molecules and the nucleosynthetic waste products of stars that just happened to be nearby when the solar system formed. Other stars and solar systems may be dominated by helium or carbon and may be more metal-rich than our own. But that fact does not belittle the feat of actually measuring the chemical composition of our own star and solar system. Why lumps of a desiccated asteroidal mud puddle should be of any use at all in this enterprise remains a mystery, but our knowledge of the abundance of elements in the solar system would be fragmentary and inaccurate if the French village of Orgueil had never been spattered with cosmic mire.

Some Suggestions for Further Reading

Allegre, C. 1992. *From Stone to Star*. Cambridge, MA: Harvard University Press. Chapter 7 of this nicely crafted little book contains a superb explanation of the origin of the elements.

Anders, E., and Grevesse, N. 1989. "Abundances of the Elements: Meteoritic and Solar." *Geochimica et Cosmochimica Acta*, vol. 53, pp. 197–214. The current bible of cosmic abundances; a technical compilation of the latest meteorite and solar measurements, challenging reading for almost everyone.

Burke, J. G. 1986. *Cosmic Debris: Meteorites in History*. Berkeley: University of California Press. An authoritative history of the science of meteoritics; Chapter 8 includes a nice summary of the evolution of thought and research on cosmic abundances.

Hutchison, R. 1983. *The Search for Our Beginning*. Oxford, England: Oxford University Press. A very interesting, nontechnical introduction to meteorites, containing information on carbonaceous chondrites and cosmic abundances.

Noyes, R. W. 1982. *The Sun, Our Star*. Cambridge, MA: Harvard University Press. Probably the most readable account of the Sun's properties and mysteries by one of the world's leading authorities; Chapter 2 includes an excellent description of solar spectroscopy and chemistry.

Suess, H. E. 1987. *Chemistry of the Solar System*. New York: John Wiley & Sons. An introduction to cosmochemistry by one of the people who invented this discipline; the book has informative sections on cosmic abundance patterns and their interpretation.

Living in the Fast Lane

────

A Planetary Foothold for the Spark of Life

My daughter's driving ambition, at least for the moment, is to be a zookeeper. She is positively fascinated by life's endless variety, which may explain why our dwelling often takes on the appearance of a menagerie. Our backyard, like the Earth itself, is an oasis where almost any biological innovation, no matter how bizarre, can find a place to live, at least for a time. This limitless diversity of critters and vegetables, however, is an optical illusion. The cat and the turtle, the newt and the hermit crab, the green slime in the fish tank and my daughter herself all can be considered one kind of life, at least from a biochemical perspective. Everything alive on our world, whether blossoming in the field, slithering through the swamp, or reading this book, is basically composed of the same chemical building blocks: proteins made from twenty amino acids, information-carrying molecules constructed from five nucleotide bases, and polysaccharides devised from a few simple sugars. Perplexingly, life on Earth uses only a tiny fraction of the millions of varied structures and chemical reactions that are possible for carbon molecules. These almost monotonous limits on biochemistry imply that zoo, forest, and pond scum must have descended from a common ancestor.

How did this ancestor, this spark of life, originate? Geology scarcely addresses this question. Paleontology, the geological sub-specialty dealing with fossils, is mostly preoccupied with detailing life's evolution rather than its beginning. Biology likewise pays little attention to the origin of life, because extant organisms provide scant clues and the problem is not amenable to experiment. For many people, even addressing the riddle of life's origin may seem more like science fiction than science. Some consider this question to be more properly in the realm of religion, and, in fact, a few have turned it into a battleground for the cause of biblical inerrantcy.

FOSSIL EVIDENCE

The Earth itself, by virtue of its geological activity, has done a pretty thorough job of erasing vestiges of its earliest inhabitants. But, against all odds, a few tantalizing tidbits of information have survived. This most ancient fossil record provides several rather important con-straints on the origin of life. First, we know that organisms came into

Petrified mounds were formed by algae that lived 3.5 billion years ago. These rocks are in western Australia. (*D. Lowe, Stanford University.*)

being very early in our planet's history. There is definite evidence for life, in the form of fossilized colonies of bacteria, in Australian and African rocks that are 3.5 billion years old. Lumps of rock resembling large cabbages with delicately laminated interiors are very similar to algal mounds found at the edges of modern seas. Contained within the ancient mounds are petrified bits of threadlike bacteria, the oldest known fossil cells. It is difficult to trace life back any farther into the geologic record, because any fossilized remains in more ancient rocks generally have been cooked and contorted beyond recognition. However, there is some indication that organisms may have already existed 3.8 billion years ago. Rocks of this age in Greenland contain tiny carbon spheres that are thought to be biological structures.

A second important contribution of paleontology to the problem of understanding life's origin is the not so surprising recognition that the earliest organisms were extremely simple. In fact, nothing more advanced than single-celled bacteria seems to have existed before

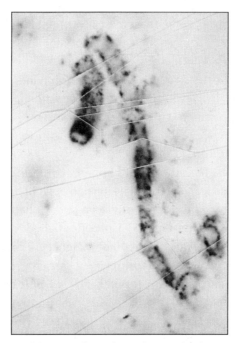

A fossilized filament of bacteria from Australian rocks that are 3.5 billion years old. These remains are the oldest known life form on Earth. *(S. Awramik, University of California, Santa Barbara.)*

about a billion years ago. From this humble beginning sprouted all of life's present glorious manifestations. Reinforcing this conclusion is a secret seeded in the genetic codes of all living organisms: Analyses of strands of DNA reveal a biochemical family tree, showing how life forms are related through branching evolution and pointing to a simple, common ancestor. What biochemistry or paleontology do not disclose are the precise steps by which the nonliving became the living, and the point at which a qualified observer would have pronounced this stuff alive.

PANSPERMIA

It was once believed that modern organisms generated themselves spontaneously from nonliving matter. Frogs miraculously arose from mud, squirming maggots popped into being from decaying meat; in short, the inanimate Earth was thought to be a place itching to be alive. The absurdity of this idea is easy to fathom from our current knowledge that frogs come from mud-dwelling tadpoles and maggots hatch from eggs laid in spoiled meat; spontaneous generation does not occur in the modern world. But what about the very first life on Earth—did it not have to arise from the nonliving? Charles Darwin believed that all living things were the evolutionary descendants of some simple, spontaneously generated life form. Did this magic just happen on its own, or did the transmogrification of nonliving into living require the touch of God?

Or perhaps this magic did not happen at all. One partly scientific and, as it turns out, partly religious alternative was championed during the last century by a prominent physicist, Sir William Thomson, later known as Lord Kelvin. In an 1871 presidential address to the British Association for the Advancement of Science, Thomson suggested that meteors (then known as aerolites) had carried germs or seeds from other worlds to our own, a proposal later to be called panspermia. Thomson's proposal for importing life was motivated by a desire to replace the godless Darwinian notion that living organisms had originated spontaneously on the Earth with a scientific view that was more compatible with scriptural demands for a Creator. Instead of a purposeful touch, the Creator sneezed on the Earth from afar. But

Thomson was sorely mistaken if he had hoped that panspermia would be embraced by either the scientific or religious establishments of his day. Most of his scientific colleagues considered the proposal a monstrous joke, and the orthodox clergy were aghast. One eminent zoologist spoke for both sides when he ridiculed Thomson's proposal as "creation by cockshy—God Almighty sitting like an idle boy at the seaside and shying aerolites (with germs) mostly missing, but sometimes hitting a planet!"

Amid all this scoffing and fuming, however, was an attentive and deadly serious Otto Hahn. Soon after Thomson's bombshell, Hahn, a professional lawyer and part-time geologist, reported his discovery of fossilized organisms in meteorites. But Hahn was not content with just microscopic germs and spores. He described the petrified remains of delicate algae and ferns in meteoritic irons, and he identified fossil corals, sponges, and crinoids in stony meteorites. These astounding biological disclosures were corroborated by a German zoologist, who even christened one prominent coral species *Hahnia meteorica* in honor of its discoverer. However, Hahn's report soon was taken to task by his scientific contemporaries. Other geologists and zoologists found no physical or chemical resemblance between the meteoritic forms claimed to be fossils and true fossil organisms. After scathing criticism from nearly every quarter, to which Hahn did not reply, the issue was put to rest, at least for a while.

Panspermia, as proposed by Thomson, rested on an earlier discovery that chondritic meteorites contained organic matter. Organic compounds are made mostly of carbon, oxygen, nitrogen, and hydrogen, all elements that are essential for life processes as we know them. Carbon, oxygen, and nitrogen are the only elements that routinely form multiple chemical bonds with one another, which result in strong but flexible structures suitable for constructing the walls of cells and the fibers of muscles. Hydrogen is an energy broker that moves electrons, the fundamental unit of energy, within and between molecules. Chemists in Thomson's day determined that the molecular structure of organic matter in meteorites was similar to that of coal, and the fact that coal formed from the decomposition of plants and animals was already well understood. It was logical enough, then, to assume that the organic matter in meteorites formed in a similar way. However, Thomson had overlooked a crucial piece of information.

More than a decade earlier, a French chemistry professor, Marcellin Berthelott, drew an important distinction between organic matter and living organisms. Berthelott had synthesized hydrocarbons from chemicals in his laboratory, and in so doing had demonstrated that life's vital force was not required for the formation of organic compounds. That meteorites contained organic matter was beyond question, but Berthelott unequivocally showed that this material did not necessitate a biologic origin.

Heated controversy about whether meteorites contained the spark of life erupted again in the 1930's, when bacteriologist Charles Lipman of the University of California at Berkeley brazenly announced that he had cultured living cells from meteorites. Lipman had sterilized the surfaces of the meteorites before placing them in cultures and so believed that he had removed any traces of terrestrial bacteria. Articles in the *New York Times* reporting "life in meteors" aroused fellow scientists, generating the same kind of invective as had Hahn's work in the previous century. One skeptic wrote that "Lipman's excursions into the field of life beyond this globe must be considered as a flight of imagination through space." When others duplicated Lipman's experiment, they found that the cells cultivated from meteorites were identical to the everyday bacterial contaminants of terrestrial laboratories. Apparently, sterilization did not completely eradicate bacteria that had worked their way into the meteorites after their arrival on Earth. Lipman at least partly recanted his assertion that the cells were extraterrestrial life forms, and there the matter ended, at least temporarily.

In 1961 the specter of panspermia resurfaced yet again at a meeting of the New York Academy of Sciences. Bartholomew Nagy and Douglas Hennessy of Fordham University and Warren Meinschein of Esso Research presented a paper comparing the chemistry of organic matter in meteorites with that of living tissue. Although they did not suggest that living organisms had been carried aboard meteorites, they did state that life must have existed on the parent bodies from which the meteorites came. Eight months later Nagy and George Claus, a microbiologist at New York University Medical Center, reported the discovery of tiny clumps of matter, which they dubbed organized elements, in meteorites. They suggested that these microscopic clumps were the recognizable fossil remnants of extraterrestrial organisms.

The organized elements were much smaller than the "fossils" of Otto Hahn, making their characterization more of a challenge. Detailed examination of them by other scientists, however, revealed that many organized elements were actually oddly shaped mineral grains, not organic matter. But a few were demonstrably from living things: There were pollen grains of ragweed and a few starch particles. Unfortunately for the hypothesis of Nagy and Claus, the organisms that produced these objects were the current denizens of Earth. The meteorite samples had obviously been contaminated by airborne dust before they were collected and probably while displayed in less-than-sterile museum cases.

The other tantalizing bit of evidence Nagy and company had used to buttress their argument that meteorites contained remnants of extraterrestrial life was based on the three-dimensional shapes of organic molecules. Like screws or bolts, many carbon-based molecules have the peculiar property of being either right- or left-handed. When viewed in a mirror, a screw with right-handed threads will appear to be threaded in the opposite direction. In a similar way, pairs of organic molecules can be mirror images. Each carbon atom in a molecule can be bonded to as many as four adjacent atoms or groups of atoms.

Drawings of tiny clumps of matter, called organized elements, found in chondritic meteorites. These strange structures were once thought to be fossils of extraterrestrial microorganisms.

Provided that each of the four attached atoms or atomic groups is different, there are two distinct ways to locate them around the central carbon atom: If we orient two otherwise identical molecules so that the atoms connected at the "head" and "foot" of each are the same, then the two atoms forming the "hands" can be interchanged, making either left- or right-handed molecules. Chemists have found that such mirror-image forms can be distinguished because they twist polarized light in opposite directions.

Organic molecules made artificially in the laboratory are normally equal mixtures of the mirror-image forms. In this case the effects of the two forms on light cancel each other out. Terrestrial life, however, almost exclusively uses the left-handed forms of organic molecules. "Left-handed" in this sense means that a molecule rotates light to the left. We have no good explanation for this strange phenomenon, except that it is probably a trait inherited from the earliest life form. Of course, we do not know that life elsewhere in the cosmos would necessarily be left-handed—it could be right-handed or even ambidextrous. However, the common ancestor from which life on Earth descended must have been a lefty. Nagy and his colleagues reported that the organic matter in meteorites rotated polarized light, implying a biologic origin. Follow-up studies of meteorites have generally found

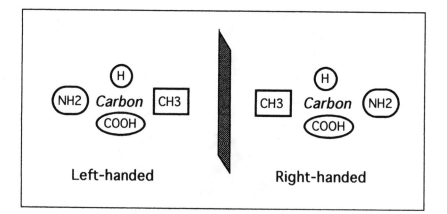

Two organic molecules can be composed of exactly the same atoms but have different geometries. The arrangement of atoms around the central carbon atom in these amino acid molecules makes them mirror images of each other. Life forms on the Earth almost exclusively use the left-handed version, as shown on the left.

no such effect, consistent with the idea that this organic matter was never part of a living organism. As with the organized elements, the optical effect Nagy reported may have been due to terrestrial contamination. A human fingerprint or a few airborne spores are all that is required to alter drastically the carbon chemistry and optical properties of a meteorite sample.

A PRIMORDIAL SOUP AND A FAINT YOUNG SUN

Despite its long and colorful history, panspermia really has little to commend it as a plausible explanation for the origin of life on Earth. A competing idea, first proposed in 1924 by Russian biochemist Aleksandr Oparin, is that life on our planet arose spontaneously from a primordial organic soup. The flavorings for this soup formed when lightning or intense solar ultraviolet radiation converted atmospheric methane and ammonia into more complex organic molecules. These molecules accumulated in the oceans, there reacting to form even more complex proteins and nucleic acids, the basic building blocks of life.

In 1953 a chemistry graduate student, following Oparin's recipe, first concocted a batch of primordial soup in the laboratory. Chef Stanley Miller constructed a kind of pressure cooker, into which he introduced mixtures of methane and ammonia gas. Steam was stirred into the mixture, while the gases were subjected to repeated electrical sparks, simulating a thunderstorm in a flask. Over a week's time the gases turned crimson. They were then cooled and condensed into a turbid broth, which was collected in a trap. Miller's experiment produced many organic molecules that are found in living tissues, a dramatic demonstration that Oparin's idea was at least possible.

Possible, yes, but is it probable? To answer this, we must know what the surface environment of the Earth was like billions of years ago. Oparin's recipe for primordial soup required that the early Earth had an atmosphere very different from the one we breathe today, one choked with methane and ammonia. It also necessitated the presence of oceans of liquid water to make the broth in which organic molecules were assembled. Both of these requirements are related to the planet's surface temperature, and here we run into a snag.

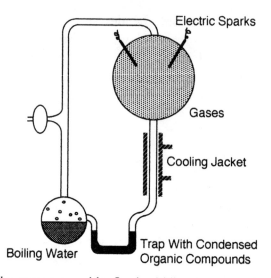

A sketch of the apparatus used by Stanley Miller to react methane, ammonia, and water vapor, forming organic molecules. The mixture was subjected to electric sparks, and then cooled and condensed in a trap.

The Earth's surface, of course, is gently roasted under the glare of the nearest star. Much has been made of the fact that our planet's orbit is conveniently placed for life, so that its temperature, like Mama Bear's porridge, is not too hot and not too cold, but just right. But has it always been so? Our understanding of stellar evolution leads to a surprising prediction—the Sun's brightness must have increased as it aged. This predicted variation in luminosity is a consequence of the gradual conversion of hydrogen to helium that occurs continuously inside the Sun. Helium is heavier than hydrogen, so the density of the Sun's interior increases as this transformation proceeds. To balance the larger gravitation force caused by increased density, the pressure and temperature of the solar interior must increase. A higher temperature, in turn, causes fusion reactions to speed up, so that the Sun produces more energy. This is not just a minor effect—the brightness of the Sun 4.5 billion years ago is estimated to have been no more than about three-quarters of its present value. Carried to its logical conclusion, the decreased solar luminosity would have resulted in a significantly lower temperature on the Earth's surface, sufficient to freeze the oceans solid for the first several billion years of the planet's history. Of course, we know that this did not happen, because sedimentary rocks from this period clearly formed under water. This apparent discrepancy between prediction and observation is sometimes referred to as the faint young Sun paradox.

There is a way out of this dilemma, though. A different atmospheric composition could have produced higher temperatures, offsetting the effect of lower solar heating. Some atmospheric gases have the ability to capture the Sun's rays as heat, through the "greenhouse" effect. Ammonia is an excellent greenhouse gas. Atmospheric ammonia would simultaneously increase surface heating to allow for liquid oceans and serve as raw material for making organic molecules. And so it seemed, for a short while at least, that the paradox was solved and Oparin's idea about the origin of life remained viable. Unfortunately, there was a catch: We now know that atmospheric ammonia, as well as methane, can exist only on a world without sunlight. The Sun's ultraviolet radiation rapidly breaks down these gases into carbon dioxide, nitrogen, and hydrogen. Many scientists now believe that an early carbon dioxide–rich atmosphere is a more plausible idea than an acrid blanket of ammonia and methane. Carbon dioxide too is a potent greenhouse gas (that is why we are so concerned now about the fact that its concentration is increasing), providing another way around the problem of a faint young Sun.

COOKING METEORITE INGREDIENTS

So the temperature problem was solved, but at the cost of Oparin's recipe. If the atmosphere on the early Earth lacked methane and ammonia, it may have been missing the very ingredients necessary to make Oparin's organic soup. Where then could the organic flavorings have come from, and how were they added to the soup?

Perhaps the soup was flavored with organic molecules from meteors or comets. After all, meteorites contain plenty of organic matter, much of it already in the form of complex molecules. Virtually all of the organic compounds necessary for the construction of proteins and other complex molecules in living forms have been found in meteorites. Numerous impacts on the early Earth must have heavily seasoned it with the spice of life. But the organic matter in meteorites clearly is not alive and never was.

After meteorites plopped into the early oceans and released their bounty of organic matter, reactions between these molecules must have occurred before the spark of life appeared. Mixtures of amino acids that are heated in the laboratory form large, proteinlike molecules. These newly concocted, complex molecules then cluster into microscopic balls that resemble simple one-celled organisms. It is possible that the earliest life forms arose in this way. The simplest of our planet's organisms are thermophilic (heat-loving) bacteria, perhaps supporting the idea that roasted chemicals led to life.

A natural source of heat sufficient to drive reactions between organic molecules is concentrated at certain locations on the floors of modern oceans. Where magma forms new ocean crust along midocean ridges, seawater seeping down through cracks and fissures is heated. This scalding water then cycles back to the surface, charging out of vents and warming the cold ocean nearby. Huddled around these vents are curious forms of marine life, the only known communities that do not depend on sunlight as a source of energy. It is not hard to imagine that similar hydrothermal vents on the sea floor may have also been the sites where beginning life forms originated billions of years ago. In fact, organic reactions still must be occurring at these locations. Perhaps newly created life forms appear from time to time around modern vents on the sea floor, although these molecules and cells would quickly be gobbled up by existing vent community organisms.

Life conjured up during this sort of chemical evolution, whether at hydrothermal vents or not, is spontaneous generation, magic really, and there is no getting around this vexing fact. Manufacturing durable organic molecules that clump together into cell-like structures is a good first step, but it is not yet life. In fact, life may be a giant leap from these gooey clots of matter. And the complex molecules produced by roasting simpler molecules are not only those used by life forms, but many others as well, organic garbage that would certainly get in the way of life-giving, metabolic reactions. One researcher has described the synthesis of organic molecules as like playing a violin, superb when well executed but appalling otherwise. It is hard to see how Mother Nature could have produced anything other than a mass of tar. Of course, it was necessary only to make one living entity, and all the rest presumably would evolve from that biologic accident.

Worried by the fact that organic synthesis is not a game for beginners and by the observation that the molecular functions of even the simplest organisms appear to be very complex, a few scientists have suggested the need for a template on which to fabricate living forms. The scaffolding they favor is a tiny mineral grain, a crystal on which carbon, hydrogen, oxygen, and nitrogen atoms could attach at specific sites and thereby react in some controlled way. A few carry this a step further (a step too far, in my opinion), proposing that crystals themselves were a sort of primitive life form. The basis for this idea is that the mineral structure, a kind of crystalline gene, carried information that allowed it not only to replicate but to compete successfully in the geologic world. In any case, the community of crystals was eventually supplanted by organic life, a parasite that grew first on the mineral template, then evolved on its own, and ultimately took over the world. One commonly suggested template is clay, a mineral whose two-dimensional layers might provide an ideal substrate on which to grow an organic life form. The notion of forming life from clay has biblical roots, although going from kaolin to Adam in one step is certainly not intended as part of this hypothesis.

SURVIVING IMPACTS

Whether chemical reactions that produced life were driven by heat or facilitated by mineral templates, the raw materials for these reactions still may have been carried in by meteorites. A potential problem, though, is that it is tricky to get organic matter in meteors to the planet's surface without being vaporized during impact. Tiny particles of infalling cosmic dust, and even fist-size meteors, are slowed by the atmosphere and make relatively soft landings. Their organic burdens remain intact. However, only a small part of the total amount of organic matter has been brought to the Earth in such small objects. Large meteors and comets arrived on our planet with a bang, impacting at velocities of twenty kilometers per second or more. In the ensuing explosions, the organic matter was likely to have been destroyed. There are several ways, though, that even large objects might have been decelerated before impact. A thicker primordial atmosphere could have provided an emergency brake for large bodies. Alternatively, large meteors might have fragmented in the atmosphere into pieces small enough to be slowed.

The proposal that meteors and comets, especially large ones, brought the organic building blocks of life to Earth has another, more ominous aspect as well. Large impacts certainly qualify as health hazards, and the Earth was pounded fiercely and often during the first half billion years of its history. Life's spark must have led a precarious existence when our planet was in the fast lane. Even modest-size asteroids would be capable of vaporizing the entire photic zone (the uppermost 200 meters) of the present-day oceans, where most marine organisms live, and such ocean-evaporating impacts may have been commonplace on the early Earth. Some organisms clustered around deep-sea vents might have survived, but their brethren that sought the warmth of sunlight were in constant danger. The earliest biological forms actually may have been reinvented a number of times as one massive impact after another exterminated the fledgling life stirring in a particular pot of organic soup.

We obviously do not yet have all the answers to the riddles of life's origin, but progress is being made. The ever-changing patterns of living organisms, and even some vague hints as to its origin, are preserved in the rock record. But fossils could never tell the whole story, because the early Earth was not a stoppered container—it was open to, and contaminated by, the cosmos. Raw materials for life, in the form of complex organic molecules, arrived piggyback on hurtling meteors and comets. The synthesis of the stuff of life did not have to start from scratch. Of course, that does not diminish the incredible fact that these imported chemicals somehow organized themselves into a self-replicating, biologic entity, the common ancestor of the strawberry, the stegosaurus, and the starfish. This primitive ancestor might have arisen when an organic soup was zapped with lightning bolts, when dissolved chemicals were roasted at hydrothermal vents on the ocean floor, when organic molecules attached themselves like parasites to bits of clay, or in any number of ways we have not even imagined. But whatever the means of life's origin, the geologic surroundings must have played a critical role. Life is part and parcel of the Earth, just one aspect of our planet's complex geologic evolution. That life found a precarious foothold on this battered world is no less a miracle than the now-discredited idea of panspermia, and perhaps no more divine than the other wondrous workings of the geologic solar system.

Some Suggestions for Further Reading

Burke, J. G. 1986.*Cosmic Debris: Meteorites in History*. Berkeley: University of California Press. A scholarly account of the history of meteoritics; Chapters 5 and 9 deal with the development of the panspermia concept.

Cairns-Smith, A. G. 1985. *Seven Clues to the Origin of Life*. New York: Cambridge University Press. This exquisitely crafted little book makes the case for inorganic crystals as the first life forms; it does not agree with the conclusions of this chapter, but it is engrossing reading.

Cloud, P. 1978. *Cosmos, Earth, and Man: A Short History of the Universe*. New Haven, CT: Yale University Press. An excellent introduction to the Earth, its place in the cosmos, and the evolution of life. Part 3 paints an especially interesting picture of the beginning of life.

Dickerson, R. E. 1978. "Chemical Evolution and the Origin of Life." *Scientific American*, vol. 239, no. 3, pp. 70–85. A very readable account of the transformation of chemicals into organisms.

Stanley, S. M. 1987. *Earth and Life through Time*. New York: Freeman. An authoritative text with an excellent chapter on the earliest era of Earth's history and the origin of life.

Wicander, R., and Monroe, J. S. 1989. *Historical Geology: Evolution of the Earth and Life through Time*. St. Paul, MN: West Publishing Company. A well-written and up-to-date college text that has an excellent chapter on ancient life.

Index